Muscles and Bones

Muscles and Bones

Charles Kovacs

Floris
Books

First published in volume form in 2006
Fifth printing 2021

 Also available as an eBook

British Library CIP Data available
ISBN 978-086315-555-0
Printed in Great Britain by Bell & Bain Ltd

 Floris Books supports sustainable forest management by
printing this book on materials made from wood that
comes from responsible sources and reclaimed material

MIX
Paper from
responsible sources
FSC® C007785

Contents

Muscles and Bones: Anatomy

Foreword

Charles Kovacs was a teacher at the Rudolf Steiner School in Edinburgh for many years. The Waldorf/Steiner schools sprang from the pedagogical ideas and insights of the Austrian philosopher Rudolf Steiner (1861–1925). The curriculum aims to awaken much more than merely the intellectual development—it seeks to educate the whole being of the growing child, that each may develop their full human and spiritual potential.

During his time as a teacher Charles Kovacs wrote extensive notes of his lessons day by day. Since then these texts have been used and appreciated by teachers in Edinburgh and other Steiner-Waldorf schools for many years. This book represents the way one teacher taught a particular group of children, other teachers will find their own way of presenting the material.

The subjects covered in this volume are taught to young people aged 12 to 14. This is an age where their consciousness is informed by a deep interest in the world around them, as well as in the physical and psychological changes they experience in themselves. The way Kovacs approaches the subject is one that would hopefully engender in the young person's soul deep feelings of wonder and reverence for the intricacies of the human body. However, it is important to remember that this particular approach, though still totally valid, was developed close to fifty years ago and should be read and interpreted as such.

Astrid Maclean

Health and the
Human Body

1

Uprightness and the Spine

What started the Age of Discovery more than anything else was the trade in spices; it was the food people ate in those days and what they put into the food, which first brought wealth and power to Venice. This gave Prince Henry the idea of getting this trade for Portugal, for his own country. If people in Europe had not had this taste for seasoned, spiced food there would have been no reason for the Portuguese to sail out into unknown seas, no school of navigation and no king would have given Columbus ships to sail across the Sea of Darkness.

And so we must now get to know something about food — what we eat and drink, why we eat and drink — and that means we must learn something about the human body as a whole. In astronomy we heard of planets and fixed stars, of the vast worlds of the cosmos, but the human body, which is so small in comparison, is just as full of wonders — just as complicated — and just as mysterious as the cosmos. But there is also a reason why knowledge of the human body is even more important for us than knowledge of the cosmos.

There was once a king who asked a wise man: "Which of all my treasures — the gold and precious stones, the palaces and castles — which do you think is most valuable?"

And the wise man answered: "O King, if you had a very bad headache, would you enjoy your treasures, your servants, your palace?"

"No" said a king.

And the wise man said: "You see, the most valuable possession for you and everybody else is health. Without health, neither riches nor power nor knowledge can be enjoyed."

Health is the most precious possession for all living things, for animals as well as for the human being. But animals don't have to learn about health or about the way their bodies work. There is kind of thrush that feeds on insects — flies, gnats, spiders. Now this thrush sometimes eats a poisonous spider which would make it ill. But whenever the bird has eaten such a poisonous spider it quickly flies to a bush with black berries and eats some of these berries. The strange thing is that the berries of this plant (called Belladonna) are also poisonous, but the poison of the berries and the poison of the spider cancel each other, and so the bird has no ill effects. The bird eats the poisonous berries only when it has eaten a poison spider, not otherwise.

The bird has no need to learn this. It knows it by instinct, but we human beings have to *learn* what is good for our body. We often do things, or eat things which are not at all good for the body; we have no instinct to warn us. And that is why we must learn what the body needs and how to take proper care of the body.

At one time or another all of you have had a bad cold, and perhaps you had a temperature. So there was something wrong with your body. But your *mind*, your spirit was also somehow not all right; if you had come to school in this state you would not have learned much, you could hardly have taken in what was said. Yet it was not your mind that was ill, it was the body. How is it that a little trouble like a cold, a slight illness of the body, can also do something to the mind? Now think of a musician — a violinist — he may be a very good violinist who can play beautifully, but if a string on his violin is broken, or even if only one string is out of tune he cannot play good music on the violin. In every one of us there is the spirit, and this spirit is the musician, and the body is the instrument of the spirit.

Of course the body is a much more complicated, a much more wonderful instrument than any violin or piano made by human hands, but it is the instrument of the spirit. But a good musician will always look after his violin well, he will treat it with great care. And in the same way we must look after the precious instrument of our body.

Now when you take violin lessons, one of the first things to learn is how to hold your violin. You would never get a good tone from a violin unless you first learn the proper way to hold it. And it is the same with the body, one of the first things we learn in life is how to "hold the body" in the right away, which means to walk upright. We learn it so early in life that none of us can remember it, but we do learn it. Have you ever watched a little toddler making his first attempts to stand up? No one told the baby, "It is time you learned to walk." No, it is the spirit in the little child that does it, the spirit which wants to hold the instrument in the right way, that is upright.

It is not the body itself that wants to stand up, it is the *spirit* that wants it. For the body it is really quite difficult to stand up — the toddler falls down a good many times before he manages to walk. It is not easy to *balance* the whole weight of the body on two legs. If you make a human figure of wax or plasticine and try to make it stand on two thin legs you will see how difficult it is. But it is not difficult at all to make wax or plasticine animals stand on four legs.

It is quite a difficult thing to balance the whole weight of the body on the soles of our feet. We can even balance it on one foot: we always do when we walk, first we are on one foot, then on the other. It only seems so easy to us because we have practised it since we were little toddlers.

But we could not balance our body on our legs if there was not one thing which keeps the body upright — the spine. If you feel the middle of your back with your fingers, you can feel the bones of your spine. Animals too have a spine, but their spine is horizontal, and the head, the ribs, the legs just hang down from the spine. We human beings carry our spine vertically, and the spine not only carries the head but also the shoulder blades, the arms, the ribs — all this weight must be carried and balanced by the spine.

The Greeks made strong pillars to carry the roof of a temple — that was easy — the temple did not move about, it always stood in the same place. But a spine as hard and stiff as a pillar would be of no use to us — we could neither bend nor turn left or right; we would be as stiff as stone pillars. But our spine is much more wonderful than a pillar. Our spine is made up of round discs, one on top of the other, and between these disc-shaped bones there are discs of cartilage. If there were no cartilage discs, the bones would jar against each other when we jump. These discs also make it possible to turn and bend our back. Sometimes one of the discs slips a little; such a slipped disc is very painful.

The bones which make up the spine are called vertebrae. It is because the spine is such a clever arrangement of vertebrae and muscles that we can walk upright but, unlike Greek pillars, we can also bow, bend and turn.

2

Posture and Walking

We saw how important the spine is for our upright walk. The spine has to support a considerable weight, it must be strong but at the same time flexible, so that we can bend and bow and turn.

When a baby is born, the vertebrae, the bones of the spine, form a straight line. As soon as the child begins to stand up and to walk, the spine curves. This slight curve is necessary. If you try to walk with your spine as stiff as a ramrod you will see how tiring this is. So the slight curve of the spine is necessary, but the curve should remain slight — it should not become a pronounced curve.

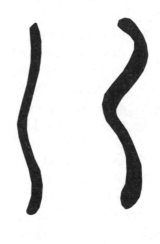

As the weight carried by the spine is quite considerable, there is always a temptation to let this weight pull on the spine and not to use your muscles to pull against the weight. And what happens then? Your head is bent forward, the shoulders become round, and the top of the curve becomes exaggerated. As soon as this happens the bottom of the curve is also exaggerated and your back at the waistline becomes hollow. And what happens in front? Your chest is drawn in, and the belly comes out.

Now the way in which we hold our bodies is called posture. The lazy, slumping posture is not only ugly, it is also unhealthy.

Inside the chest there are the lungs and the heart, and when the chest is pulled in, the lungs and heart are cramped, and in the long run, they suffer from it. And in the belly are the stomach and the intestines, which digest the food you eat, and if the belly is kept in this slack position then the stomach and intestines sag forward and cannot work properly. So a slumping, lazy posture is unhealthy as well as ugly.

You must also think of the right posture when you are sitting, not only when you are standing or walking. The lower part of your back should be against the back of the chair. When you bend forward to read or to write you should bend from your hips, not from the waist, not from the shoulders.

The golden rule to keep a good healthy posture is: Stand tall, walk tall, sit tall.

Think of the women in India who walk with a queenly grace because they were used to balancing water jars on their heads.

To have the right posture not only depends on our spine, but also on our legs and feet. As we are upright beings, the whole weight of the body presses down on the soles of our feet. But our feet are not squat like an elephant's feet; they are a strong and at the same time a delicate structure. The weight of the body is not on the whole of the foot but only on three points: the toes, the ball, and the heel, so the arch between the ball and the heel is left free. And when we walk the weight comes down first on the heel, then we go forward to the ball and lastly to the toes. It is really a rolling motion. That is why we don't stamp or trample like elephants but have our own human gait.

Now these three points which carry the weight — toes, balls, heels — have especially strong bones. The strongest bone is in the heel, but where the arch is, the bones are not particularly strong, they are not meant to take the weight. And when a person has fallen arches or flat feet the curve of the arch is flattened and the weight of the body presses down on bones which are not very strong, and so the foot easily gets tired.

Some people who live in a hot climate walk barefoot, but in our climate we need shoes. And it is most important, especially when we are still growing, that we wear the right kind of shoe. A well fitting shoe should have a little extra length beyond the big toe, but it should not have extra width, it should be as wide as the widest part of the foot. At the heel the shoe should fit closely: there should be neither rubbing nor slipping.

Ill-fitting shoes give rise to corns (hard skin) or to even more painful bunions (a bony enlargement of the joint of the big toe). They are signs that the feet are protesting against ill treatment. It seems impossible that people in their right mind would deliberately hurt themselves, but a good many people do just that for the sake of some silly fashion.

There are pointed shoes. The toes of the foot do not go naturally into a point, and if the foot is put into a pointed shoe, the toes are crowded inwards and in time become crooked. If the shoe is too small, the toes are not only crushed together but also pushed backwards resulting in corns and bunions.

If you really want to harm your foot and your whole body, then wear high-heeled shoes. The foot is distorted; it is not in its natural position. And the whole weight of the body falls on balls and toes through the ankles, while the foot's strongest bone in the heel hardly carries any weight. And so the front part of the foot is overworked, overstrained while other parts, which should work, are not working at all. If you wear high-heeled shoes the muscles in the calf of your leg get slack and weak, and they become shorter. And then you cannot easily change back to normal shoes again.

But it is not only the foot and the leg which suffer from high heels. If you stand on such high heels, your whole body is pitched forward and is thrown out of the vertical. And then, to keep the body from falling forward, your muscles pull the body back, which makes the spine curve more than it should, it makes a hump at the back of the neck and an ugly hollow in the back at the waist. It produces the ugly and unhealthy posture mentioned above. High-heeled shoes can cause headaches, bad blood circulation, and all kinds of other complaints.

A shoe must help you to walk with ease and comfort, and not make you strut or slouch. And walking itself is not just a means to get from one place to another, walking is a health-giving activity. People who sit all day in some office, then sit again in cars and buses, then sit again when they come home, are harming their bodies; the heart, the blood circulation, the digestion, all get lazy and sluggish. Our body needs walking just as it needs air and food.

But walking must be in the right way: the feet should be pointing forward and the ankles not too far apart. Walking slowly along the chalk line on the floor, our feet should be parallel to the line and not at angles.

Walking in the right way and the wrong way

Watch other people's walk, for you can learn a lot about a person from the way they walk. A happy person walks light-footed, a sad man drags his feet. People who have self-confidence walk with a firm step, people who are lazy slouch.

3

Head, Trunk and Limbs

We have heard about our upright walk, how important it is that we walk, stand and sit properly. But now I want to look at the result of our walking upright.

Just think how firmly a horse or a cow stands on the ground with its four legs — it does not have to balance. Of course, a horse can rear up on its hind legs, but it cannot keep this posture for long. Yet that is exactly what we do all the time; we keep our body balanced on two legs, and they are rather slim and slender. But because the task of carrying the body is given to our legs, our arms and hands are free to do the thousand and one things we human beings can do.

If a lioness has killed a sheep and wants to bring it to her cubs, she carries it in her mouth. If an elephant wants to pluck a branch from a tree, it uses its trunk. If a bird wants to build a nest, it uses its beak. The animals use their head, or a part of their head, as we use our hands. And so for the animal the head is a kind of limb, it is used as we use our hands.

The human head does not have to do much physical work, but it can think, have ideas, invent new things. The animals never invent anything new. The cells in a beehive, the nests of birds, they have remained the same for millions of years. Only the human head can think out new things because it has no physical work to do. And only human hands can write, draw, paint, make tools and machines which run faster than horses, fly higher than birds and swim faster than fish. And all this — that the hands are free to make things, that the head is free to think — comes from our upright walk.

When you learn to play the violin, you learn first to hold it properly. And the spirit that lives in every human being learns first to hold its instrument, the body, properly, that is upright. But just think what a wonderful instrument this body is: the head is used for thinking, the hands for making things, and the legs for carrying the body. But the arms and legs do have something in common: with our arms and legs we do things; they are the instruments of doing. The head is the instrument of thinking. And in between — our heart, our lungs — all this middle part is the instrument of feeling.

Our body is really three instruments in one. And we can see the three different instruments of the spirit in the form and shape of the body. Each instrument has its own particular shape.

The head of course is round; it is a sphere, like the blue dome of the sky above us. The bones of the head form a *sphere*.

The limbs — arms, hands, legs, feet — are quite different. The bones of the arm, the fingers are straight — of course not like a ruler, but compared with the rounded head, the limbs are straight. They are really tubes, for the bones are hollow inside.

The sphere of the head is a picture of the cosmos, of the cosmic sphere. The straight lines of the limbs belong to the earth.

But between the spherical bones of the head and the straight lines of the limbs there are the ribs. If you press your fingers below the arms, you can feel the rib bones. And the shape of the rib bones is something between the sphere of the head and the straight line of the limbs. They are round, but they are not completely closed. They form a kind of half-open cage and this part is often called the ribcage.

If you make a fist of your hand, it is like your head bones. Hold your hand wide open, it is like the limb bones. And hold it in an in-between position, and it is like the rib bones. So the shapes of your bones tell us that our body is really a combination of three instruments.

There is something else to show how different the three instruments are. Inside the head is the brain, and outside are the strong round bones which protect the soft brain. The arms and legs are just the opposite. Outside is soft flesh — the muscles — and only inside the flesh are the bones. In the middle, things are mixed: the ribcage is not as hard as the head bones, it is also a half-open cage, and just under the ribcage the body is soft, you don't feel any bones. So there are again two opposites. Head: hard outside, soft inside. Limbs: soft outside, hard inside. And the middle part is a mixture of both.

But now consider how these three instruments work. If you shake your head wildly, moving and turning it, it does not help your thinking. The head can do its work only if it is at rest and does not move. The head is meant to move very little, to rest from all physical work, so that it can think.

With the limbs it is just the opposite: they only do their task when they move. Of course they also need rest, but when the legs and arms are resting they are not doing their task. When they are doing their task, arms and legs, hands and feet are moving. And in-between the heart and lungs are not at rest like the head, they are not moving about like the limbs. But they do move: they move rhythmically — the lungs breathe in and out, the heart beats.

So thinking, feeling and doing each has its own instrument in the body:

Thinking: Spherical bones, head, at rest.

Feeling: Curved bones, heart and lungs, rhythmical movement.

Doing or will: Straight bones, limbs, free movement.

4

Sleep

Before we go on with the uprightness of the human body we must consider something else. We are not always upright, for when we sleep we are horizontal.

If asked why we sleep, we would probably say, "because we are tired." But a little baby only a few days old has not done any work, has not played, run or walked — it has not done anything that could make it tired, but it sleeps most of the time. Or think of people so ill that they are in bed for days or even weeks. They too are neither working nor walking, but they sleep more than healthy people.

So we need sleep, just as we need air or food. But we still have to find out why we sleep. When we sleep, our head does not do any work, our arms and legs don't do any work; they all rest. But our lungs and heart go on day and night without stopping. Other parts of our body may sleep and rest, but our heart knows no rest as long as we live. That is rather strange; how can the heart and the lungs go on working for seventy, eighty or ninety years without resting? Don't they get tired? The reason they do not get tired is that the heart and lungs move in a rhythm, and doing things in rhythm does not make you tired, it gives strength.

When soldiers have to go a long way they march; they walk in a steady rhythm and often have drums or music to make the marching more rhythmical. They can go on much longer and get less tired than if they walked or shuffled along any old how. Rhythm gives strength. For the same reason sailors on the old ships sang sea shanties and worked in time with the song, in rhythm.

The heart has a rhythm, the breathing of the lungs has a rhythm. But our body as a whole also has a rhythm, a rhythm which keeps the body healthy and strong: and that is the rhythm of sleeping and waking.

Just as the lungs go from in-breathing to out-breathing, so the whole body goes from waking to sleeping. We breathe in and out about 25 920 times a day (18 breaths per minute × 60 × 24), and we go from waking to sleeping about 25 920 times in an average life of about 71 years. It is a slower rhythm than breathing, but a rhythm.

Waking and sleeping are the great rhythms of life, just as in-breathing and out-breathing are the short rhythms of life. When we feel sleepy, the body wants to go from one half of the rhythm to the other half of the rhythm, and this wish — this need of the body, this longing of the body for the next part of the rhythm — makes you feel tired or sleepy. You do not fall asleep because you are tired — quite the opposite, you feel tired because the body longs for sleep, for the other half of its rhythm. But the part of your body which is already in a rhythm, your lungs and your heart, goes on just the same with its own rhythm.

But do not think that nothing happens while you are asleep. Scientists have made exact measurements and have found that children grow mainly during the hours of sleep and hardly when awake. Now we can understand why little babies sleep such a lot: because they grow very quickly, and to grow they need sleep. Of course, this does not mean you grow taller if you sleep more.

Another thing scientists found was that serious wounds heal mainly during sleep, not in the daytime. That is the reason why people need much rest and sleep after an operation. They do not need sleep because they are tired — they have not done any work. They need sleep to heal the wounds of the operation. This is true for all illnesses. Since ancient times it has been known that no medicine works without sleep. The greatest healer of all is sleep.

As long as children are still growing, they need sleep much more than adults who don't grow any longer. If the body has to grow as much in six hours as it would like to grow in ten hours, then it will not be built up as strong and healthy as it should be — the body has to rush and a rushed job is never a job well done. So you only have to ask yourself: "Do I want a healthy body when I grow up, or a weak, unhealthy body?" And if you want a healthy body, then you need ten to twelve hours. Adults can manage on six to nine hours — and old people don't need more than five to seven hours sleep.

Some scientists wanted to find out what happens if people do not go to sleep but stay awake as long as possible. Some of their students offered themselves for the experiment. For thirty hours the students were quite normal, but after this their minds began to show the strain: they lost their memory, they could not even remember their own names. After sixty hours — two and a half days — they began to see things which were not there. They saw enemies attacking them and cried out in fear. At this stage the experiment was stopped. The students had a long sleep and when they awoke they were completely restored. So it is not only the body but also the mind which needs sleep. The mind and the body are refreshed and restored by sleep, and that is why we must have the right amount of sleep.

While we sleep the heart keeps on beating, but it beats slower than when we are awake; the breathing too is slower. And the temperature of our body is also lower when we sleep. That is the reason why the body should be well covered when we sleep. Another thing — it sounds strange — is that when we are asleep there is more blood in our legs and feet, and less blood in the head. And therefore the legs and feet should be especially well covered. People falling asleep in ice and snow die of the cold!

Another important rule is that one should not read or watch anything exciting or thrilling before going to sleep; the excitement keeps away the health-giving sleep and it becomes a restless sleep. Especially growing children will pay years later as adults for the nights when they did not have the deep healthy sleep they needed. Nor is it good to go to sleep after a heavy meal. Digesting food is work for the body and the food remains undigested in the stomach and prevents deep, healthy sleep, or even gives us nightmares.

The best preparation for sleep is to feel at peace, and the best way to feel this inner peace is a really heartfelt prayer to God who is the Lord of Peace. Nowadays many people suffer from sleeplessness, insomnia, and take all kinds of drugs and pills. This is very bad for the body. The real cure for insomnia is quite different. Sleep, as we all know, is one half of a rhythm. And if people suffering from insomnia get into the habit of always doing certain things (no matter what it is) at the same time every day, then they bring rhythm into their life, and sleep will come without drugs.

Awakening is also important. It is not good at all to be awakened suddenly by the shrill noise of an alarm clock. One should awaken gradually, like the sun rising. If you tell yourself, "I will wake up at 7.30," you will wake up without a clock, provided you had your ten to twelve hours sleep.

5

The Skin

We have seen that for the three different parts of the human body — the head, the limbs, and the middle part between — we can "imitate" the three structures with our hands. If we make a fist, the hand is rounded like the head, if we hold the hand open, the fingers are straight like the limbs, and if we hold the hand in between it is like the ribs. What happens when we throw a ball? We first pick up the ball or catch it, and to hold the ball or to pick it up, our hands must become rounded like the head. But to throw the ball far we move our arm and open our hand wide so that it becomes like the limbs. To receive, the hand becomes like the head, and to send, the hand becomes like the limbs.

From this you can see how the head works. The head receives or takes in; through eyes, ears, we take in what comes from the world outside. Our limbs work into the world: we make things with our hands, we carry our body from one place to another with our legs. And in the middle, both things come together: we breathe in, we take something in; we breathe out, we send something out into the world. Our whole life is like this: taking in and giving out. When we are young, we do more taking in from parents and teachers, while when we are older we give out to our own children, to other people.

But now we come back to the human body. What is the borderline between inside and outside? The skin. It encloses and surrounds the whole body and is something quite wonderful. In some ways the skin "takes in" like the head, for how do we know that something is smooth or rough, hot or cold? Through the skin. Just as the eyes tell us the shape and colour of a thing, so the skin tells us whether it is hard or soft, smooth or rough, hot or cold. But our skin is also like the limbs; it gives out into the world as we shall see later.

First we must see what the skin does for the body. The most important is that the skin protects the body; the skin is the body's armour. But this armour of the body is not stiff and hard like a knight's armour, it is elastic and stretches. The skin is especially elastic on our knees and elbows, which bend and stretch a lot.

Another remarkable thing about the skin is that it is waterproof in one direction. When we go out in the rain or take a bath or swim, very little water from outside gets into our body. The skin is waterproof and does not let much outside water come in. But when you sweat the skin lets water come through to the outside, and then it is not waterproof. Water is let through from inside to outside, but not from outside to inside.

But the skin is also continually changing, and to understand these changes we will now consider one particular part of the skin. It is only a small part, but we can learn a lot from it. The fingernails are part of the skin and grow from it. And they keep on growing, which is why you have to cut them. If you don't cut them, they will just break off when they get too long. But when you cut your nails you can only cut the top part, you cannot cut them further down as it would hurt. The top part which breaks off or is cut, is really dead, that is why you can cut it off without feeling anything. But lower down the nail is still alive, it is part of the living skin and it would hurt you to cut it. If you make a mark on a nail after a few days you would see that the mark has come higher, and a few days later still higher, and after a time the mark is on the dead part of the nail which you can cut off. There is always new living nail growing from below and a dead part on top.

What happens with your nail — it grows continually from the living part to an old dead part — is the same that happens with your whole skin. We continually grow new living skin below and on top there is a layer of dead skin, the topmost part of this dead skin is rubbed off. When you wash you can see little scales coming off your skin. These tiny scales are the topmost part of the layer of dead skin. So all over the body the skin is doing the same thing as your fingernail. New living skin is formed below, and on top the old skin dies and is shed, just as the top of the nail. The lower, living part of the skin is called dermis and the top part is called epidermis.

This dead layer, the epidermis, is very important for the living dermis underneath. When you cut yourself and the wound has healed, the new skin which has been formed is very tender — if something knocks against it, it hurts. But after a time a layer of dead epidermis has been formed on top and then the spot no longer hurts because the tender skin below is protected by a layer of epidermis. The epidermis is really an armour of little transparent cells which protects the living dermis underneath. If you only had the dermis, your whole skin would be like the sore spot where a wound has just healed: your skin would be far too sensitive. The epidermis protects the sensitive dermis below. The top of this epidermis, the top layer of these transparent scales, is continually "flaking off." On the inside of your hands and on the soles of your feet there is a continual rubbing off of these little scales. By doing things with your hands, by walking, you are rubbing the epidermis of hands and feet much more than on any other part of the body, and the epidermis of the palms and soles would soon be worn through. It is not worn through because we grow an especially thick layer of epidermis there. People who walk barefoot or who use their hands for rough work grow an extra thick epidermis on the soles or palms. You can tell from the extra hard skin on a man's hand that he is used to hard rough work with his hands. So the body sees to it that the tender dermis underneath is always well protected. But what is now the tender dermis underneath will in a few weeks time be the dead epidermis and will protect the new skin that comes up from below.

There is another thing which is a kind of epidermis: that is your hair. Of a single hair only the root in the skin is alive. If somebody pulls your hair, you can feel the roots hurt. But the rest of the hair, the whole length of the hair, can be cut and you don't feel anything; it is not alive. Hair grows exactly like the skin grows. At the root new living hair is formed and the old part is pushed out. Nails, skin and hair all grow in the same way.

Between the hair roots there is a proper epidermis which also flakes off in little scales. We don't wash our head as often as our hands, but it does need washing to get rid of these little scales on the scalp which are called dandruff.

Just as snakes cast off — or slough off — their whole skin from time to time, we do the same, but we do it in bits, in little scales.

Care of the Skin

We looked at the two parts of the skin: the upper part, the dead epidermis, and below the living dermis. Now this living lower part of the skin gives us our complexion. Whether we have rosy cheeks, a pale skin or a sallow, yellowish complexion comes from the dermis. Of course everybody would like to have a good complexion and rosy skin, and would like to keep it. So let us see how we can help our complexion.

Even a pale skin can get a rosy glow for a little while if you rub it vigorously with a towel. It does not last, but it can give us some idea what makes a rosy complexion. In the living dermis there are many tiny blood vessels, and if you rub the skin, the blood in these tiny vessels moves with more strength and speed. This vigorous movement of the blood gives the skin a healthy, rosy colour.

Of course, rubbing from outside helps only for a moment or two, and so we must find better, more permanent ways to make the blood move vigorously. The rules for this are quite simple: keep the whole body healthy:
— Plenty of exercise in fresh air
— Fresh fruit and vegetables
— Regular hours of sleep and enough sleep
— See that your bowels move regularly.

These are the things which the whole body needs, and if you look after these things, the skin will look after itself. Just as a thermometer tells you the temperature in a room, so the skin tells you whether you live a healthy life or not.

The health of the body works on the skin and gives you a healthy complexion. But things also work the other way round.

The skin also does something for the health of the body. We have all heard of vitamins, substances of which the body needs only very little, but these tiny amounts are as necessary for the body as the bread we eat. There are all kinds of vitamins — they are called by letters A, B, C, D and so on — and we get most of them in fresh fruit and vegetables.

But there is one vitamin which the skin itself makes and gives to the body; it is called vitamin D. The body needs this vitamin D, but it does not get enough of it from what we eat. It also gets it from our own skin. But the skin can only produce this important vitamin D if it gets enough sunshine. It needs sunlight to produce this vitamin. If someone does not get enough sunlight on their skin by living in a shady place, for instance, the skin cannot produce this vitamin D. What happens then?

The part of the body which needs vitamin D most is the bones. If they cannot get vitamin D from the skin, they become soft and the legs cannot carry the weight of the body, bending under the weight. This disease is called rickets. Many years ago when factory workers and their families lived crowded together in narrow, grimy streets where one could hardly ever see a ray of sunlight, thousands of them suffered from rickets. Nowadays we know better; we know that the human skin needs sunlight to build up strong bones, and we build our streets and houses accordingly.

One can also take in some of this vitamin D by eating milk, cheese, butter, eggs, fish oil, but it is difficult to get enough from our food, so it is important to get it also through sunlight. The Inuit, who had to live without sunshine for many months, got their vitamin D by eating raw meat.

The skin needs sunshine to make this vitamin which the body needs. But the skin only wants a certain amount of sunlight, it does not want too much of it. And so when the dermis has had as much sunlight as it can stand and wants no more of it, it makes a kind of sunshade for itself. It produces a dark colour, a dark pigment called melanin. The skin covers itself against too much sunlight with a suntan. Peoples that have lived for thousands of years in a climate of glaring hot sunshine — like the people of Africa — get this dark protection of their skin soon after they are born.

Light-skinned people have to be careful about getting a sun-tan. If a person is fair-haired or red-haired, it means that the skin is not able to make much pigment. Fair-haired people are short of pigment in their skin, and do not tan easily and should not be too long in the sun. Even dark haired people must be careful, and give their skin a chance to make its own protection slowly and gradually. If the dermis is exposed to too much sun-light too suddenly, it can be seriously damaged.

The skin is a very important and a very delicate part of the body. Being such a delicate part, it needs looking after. First of all that means we must keep it clean. We all know that, but why does the skin have to be kept clean?

The epidermis, the top part of our skin, is dead. It would become stiff and crack, but the living skin, the dermis, sends a fine oil up through the pores. This oil keeps the epidermis sup-ple and elastic.

The pores also send sweat up to the surface. Sweat has a salt taste — it is water and salt. The water of the sweat evaporates, but the salt remains. Then this salt, the oil, the dust and grime from outside and the little dead scales all mix on the surface of the skin forming a kind of mud. That's what we call dirt.

The dirt clogs up the pores of the skin stopping new oil and sweat, which is bad for the body. Also dirt is a breeding ground for germs. So dirt is not only ugly but also unhealthy, it is an invitation to germs.

Another thing which is nearly as bad for the skin as dirt, is the use of certain kinds of makeup and face creams. These creams are often a mixture of grease and wax. The grease forces its way into the pores and blocks them up, and the wax covers the openings and forms a coating over them. As a result the oil and sweat are dammed up in the pores, the skin cannot work properly becoming thin and wrinkled sooner than normal. A good complexion does not need makeup or creams, and a bad complexion will only get worse by the use of cosmetics.

All the skin really needs is a good cleaning with soap and warm (not hot, not cold) water. Then rinse the soap with cold or cool water — and you have done all your skin really needs.

The Four Elements
and the Body

Going for a walk in the hills, you might come to a pond or a lake with a large rock on the shore. On the rock some people have made a bonfire, and there is a fresh wind driving clouds across the sky and murmuring through the trees by the lake. With the rock, the lake, the wind in the trees, the flames, we have a picture of the four elements: the solid earth, the rippling, liquid water, the air moving the trees and driving the clouds, and in the midst of it all, the flames of fire. It is a picture of earth, water, air, and heat or fire; of the four elements.

In such a picture you have the four elements spread out in nature before you. Now imagine walking in the street and bumping into somebody. You might think you have bumped into something as solid as a rock, but you are mistaken. The human body is only partly solid; the body is built up of all four elements — the solid, the liquid, the air and the heat. What you see outside in nature, can also be found inside the human body. And there it is put together in a very particular way.

Let us start with the heat. Our body has a certain temperature (37°C or 98.6°F) which changes a little — a bit more in the morning, a bit less at night, a bit more when we run, a bit less when we sit. But on the whole our body has a certain temperature, a certain warmth. What is the shape of this warmth of the body? It is, of course, the same shape as the body itself. And this warmth goes right through our whole body — through our blood, our lungs, our heart. They all have this warmth, with slight variations. We have a complete "warmth body."

Now we come to the air — we breathe in and breathe out. When we breathe in the air not only goes into our lungs, but our blood carries the air into every tiniest part of our body and no part of our body could live if it would not receive air. We not only breathe with our lungs, but with every part of our body. And the air we breathe out has come from every part of our body, even the bones. The air we breathe in goes to every part of the body. It is spread out over the whole body, and so we also have inside us a whole body of air. We have this "air body" even after we have breathed out, for the body never breathes out all the air, and so always have a body of air inside us, just like the body of warmth.

And now we come to water. But let us first look at the whole earth. If you look at a globe, you will notice that there is quite a lot of blue on it; that there are really more oceans and seas than there is dry land. In fact about two thirds of the whole surface of the earth is water and only one third is dry land. It is exactly the same with the human body: only one third of the body is solid, and two thirds are water. Most of our blood is of course water, and even our flesh, our muscles are three quarters water. Only the bones and teeth are solid, but even they contain water. A fifth of our bones is water. Even the shiny part of our teeth, the enamel, contains 2% water. No part of the body is without water.

We not only have a warmth body and an air body, but also a water body. And what is more, this water body is two thirds of our body. Only one third of our weight is solid and two thirds are water.

The surface of the earth is two-thirds water — oceans, seas — but these are not fresh water, they are salt water. Similarly the water in our body is salt water. If you lick tears, sweat, or a drop of blood when you cut yourself, they all taste salt, because the water in us is like sea-water. Now we can understand why scientists think that life began in the sea, and when the first creatures left the sea to live on land, they carried the sea with them in their blood.

Two thirds of an adult's body consists of water. But now compare a baby with a very old person. The old man or woman looks "dried up," their bones are dry and brittle. By contrast, a little baby is round and soft, the bones are still soft containing much more water — it consists of three quarters water. Also, at first the baby can take only liquid nourishment. Human life begins with being more "watery," and getting old is a kind of drying up. But even in an adult weighing 65 kg, about 44 kg are water — more than 40 litres.★

We are two thirds water, salt water. We are really a "walking sea" and the solid parts, the bones, are only continents and islands in the sea. But we are only *alive* in this watery part of the body. The liquid part of the body carries the food we eat to all parts of the body, it carries the air we breathe to all parts of the body, it keeps us alive. The solid part, the bones, could not live without the watery part, just as the earth would have no life without water.

We have a warmth body, we have an air body, we have a water body, and we also have a solid body. But these four bodies are not separate, they are inside one another, interpenetrating each other in every part of the body. Take your skin: it has warmth, it has air (the blood brings it air), it has water (otherwise it would be hard) and it also has something solid (it does not flow away as water would).

And so it is with every part of the body. Throughout the whole body the warmth, the air, the water, the solid — all four elements — are inside one another. In the blood there is more water and only one fifth of that is solid while in the bones there is more solid and only one fifth water, the warmth and air are more or less the same all over the body. Everywhere there are the four elements. What you see outside separately as rock, lake, wind and fire, is together in your body as four bodies inside one another, interpenetrating each other.

Now let us look at one of these bodies: the air body. When I breathe in, the air goes to all parts of the body, it mixes into the air body. Then I breathe out, some (not all) of that air body has gone. Then I breathe in again and have new air, and I breathe out old air. So I partly change my air body every time I breathe.

★ A 10 stone (140 lb) body contains $6^{1}/_{2}$ stone (93 lb) water, almost 10 imperial gallons (13 US gal).

I have an air body, but the air of which this body is made is new air, different air, every few minutes.

But it is the same with the warmth body. The warmth of the body goes out into the air around us; on a cool day the warmth coming from our bodies makes the room much warmer than it would be if it stood empty. The warmth of our body continually goes away and new warmth made by the body takes its place. The body makes new warmth from the food we eat, and we eat more in cold weather than in hot weather. The body uses up more food when it is cold. So our warmth body too changes. We have a warmth body, but the warmth is new warmth every few hours.

And the water too changes. We drink water — tea and milk are also mostly water — and a good deal of what we call solid food, like meat and vegetables, also contains quite a lot of water. And we also get rid of water: first in our breath, for we don't breathe out only air but also water. On a cold winter's day you can see it going out as "steam" from your mouth, in warm weather you only have to breathe against cold glass to see it. More water goes as sweat, and more as urine. So the water of our body is also constantly changed — not so quickly as air and warmth — but scientists have worked out that after about three weeks most of the water in us has been changed for new water.

The air body is changed for a new air body every few minutes. The warmth body is changed for a new warmth body every few hours. The water body is changed for a new water body every few weeks. And the solid body, the earth body too is changed.

The Changing Solid Body

There are quite a number of situations in which you can see a rainbow. Sometimes you see a rainbow in the spray of a waterfall or a garden hose, or you see a rainbow when there is sunshine where you yourself are, but a mile or so away, opposite the sun, rain is falling. The rain is both the prism which breaks or refracts the light, as well as the "screen" on which you see the rainbow.

Similarly when you look at a cumulus cloud, it seems to stand still, keeping the same shape, but all the time the tiny droplets of which the cloud is made change. At the edges the droplets evaporate, becoming invisible, and new droplets condense below. The cloud more or less keeps its shape and place, but the drops are changing, some evaporating and others condensing in their place.

The rainbow only seems to be the same, but it is formed of drops which are different from moment to moment. The cloud only seems to be the same, but it is formed of drops which are different in any moment. And it is the same with the human body: the air body is formed by different air every few minutes, the warmth body is formed by different warmth every few hours, the water body is formed by different water every few weeks. What about the solid body, the earth body? Of course every part, every organ of your body, even the blood, contains something solid. Just as the water is varied with a little more here, a little less there, so the solid parts are varied — more in the bones, less in the blood. But everywhere in the body is some solid part.

Remember the skin: constantly new skin is formed within

and the older skin is pushed outwards as epidermis and then flakes off. There is always a new nail growing, and the old nail is pushed out until you cut it off. And the same is happening in every part of your body: the heart, lungs, the blood, and even the bones are continuously built up anew and the old matter is cast away. Some leaves the body when you sweat, some in the urine, some through the bowels, and some as epidermis.

What makes your bones hard is a substance called calcium. For instance milk contains a lot of calcium to build strong, hard bones. But if people cannot get food with enough calcium in it, then after a time, there will be less and less calcium in the bones, they become brittle and thin and break easily (this sometimes happened in wartime when milk, cheese and butter became scarce). So even the bones don't keep the matter they are made of — they slowly excrete, or push out, old matter, and if they cannot get the new matter to rebuild themselves, they become thinner and weaker. But with the right food they can rebuild themselves.

So the bones, heart, lungs, skin, and blood are all like the cloud: they keep the same shape but the matter they are made of is continually renewed. But this change of the solid parts takes much longer than the change of the heat body, air body and water body. It takes about seven years.

You have a new air body every few minutes, a new heat body every few hours, a new water body every few weeks, and a new solid, earth body every few years. Taken as a whole, the body is like a cloud keeping the same shape but all the substances of which it is made are continuously changing. And after seven years hardly the tiniest particle of your body is the same as it was seven years ago.

Just think: the body you have now is not the same body you had seven years ago. There is nothing in the body today that was there seven years ago, yet you are the same person. And in seven years time nothing of the body you have now will be left, but you will still be the same person. This is what the word "soul" means — it is what makes you the same person, although the matter in your body changes from second to second, week to week, year to year.

But your soul also changes. The things you liked when you were six years old are not the same as the things you like now. And in fifteen or twenty years' time, what you think, what you feel, and what you want to do, will be quite different from what they are now. Yet you will still be the same person. What stays the same — that which you call "I" now and will call "I" as long as you live — that is what the word "spirit" means.

What you can see and touch of the human body is changing, it is coming and going like the breath (only much slower), and the thing which lasts is what you cannot see, is soul and spirit.

Regulating Warmth

Our body has a more or less constant temperature of 37°C (98.6°F). Birds have a higher temperature, and so have horses and cows. There are animals whose blood has no warmth of its own, "cold-blooded" animals. For instance, lizards, tortoises or snakes are cold-blooded. Their blood has no warmth of its own. But they do get warmth from the sunlight and love to sun themselves. During the summer they get warmth from the sun; you could say they get their warmth body from outside.

As long as there is some warmth and light from the sun, lizards and snakes can move very quickly, but towards autumn when it gets cooler, their movements become slower and sluggish. When it gets still colder, they find themselves holes, stop moving altogether and hibernate.

It is not only lizards and snakes that cannot move without warmth, no other animal nor we human beings can move without warmth. This is also true of the whole of nature: without the warmth of the sun, water would not rise from the oceans, there would be no rain and no rivers; without the warmth of the sun, warm air would not rise and there would be no wind. There is no movement without warmth.

Lizards, snakes and tortoises can only move and use their muscles as long as they get the warmth they need from the sun outside. When the sun is too low in the sky to give them enough warmth, they have to stop moving and must hibernate. But we human beings — and the warm-blooded animals — have our own "sun" in our warm blood, and so we are not dependent on the outside temperature. If we were like the lizards, we could neither work nor play on a cool day. But we can do whatever we want to do in cold or in warm weather because we have our own "sun," our own warmth body.

There are three things which the soul does: thinking, feeling and willing or doing. Where is the will, the doing part of the soul? It is in the warmth of the body. When you do something — anything that needs effort — then the soul works through the warmth. Even when you make a real effort in thinking, when you really think hard, then you can feel the warmth in your head and on your face. This warmth body is our own "sun" — it makes it possible for us to do things, to use our will power whenever we want it without being dependent on the sun outside.

And now we come back to the skin because it has a very important part to play to protect the warmth body. For the health of the body it is most important that the temperature of 37°C (98.6°F) does not change much. It is the task of the dermis — the living skin under the epidermis — to keep the temperature of the body at the same level.

When it is very hot outside, then the heat could also raise our own temperature, our own blood would get warmer which would be very bad for the heart and other organs. But even on the hottest day our blood does not get warmer, because we sweat. When we wet our hands and let them dry in the air, our hands feel cooler. When we come out of the water after a swim our wet body shivers (unless the air is very warm), but if we dry ourselves we don't feel so cold. The reason is that whenever water evaporates it takes heat away.

And when we get hot, either because it is hot outside or because we have been running, then the sweat glands in the epidermis send sweat to the surface of the skin through the pores. The sweat evaporates and takes heat away and our body temperature is kept normal. The skin sweats to protect the warmth body against getting too hot.

Once in the city of Florence in the Middle Ages, there was a great festival for which people dressed up in all kinds of fancy dress. One young man had the bright idea of covering himself from head to foot completely with a layer of gold dust. He was a splendid sight walking about like a golden statue, but in the middle of the festival he suddenly collapsed and died. His skin, smothered with gold dust, had not been able to sweat and he died of overheating.

Against cold, the protection is quite different. Long distance swimmers cover their bodies with grease so as not to get too cold when they are in the water for hours. Grease — fat of any kind — keeps the warmth in and the cold out. Fat is an insulator, a bad conductor of heat. For the same reason animals which live in the Arctic, like whales, seals and polar bears, have thick layers of fat on their bodies. Fat is like a wall that protects the inner warmth against the cold outside. And under the dermis our skin has a layer of fat to hold off cold that comes from outside. Fat people sweat more than others in hot weather, but they can stand cold weather better. Lean people sweat less on a hot day, but suffer more from cold weather.

Warmth and Clothing

We have seen how the skin protects the warmth body, how it ensures that our body temperature does not go much higher or lower than is good for us. It is like a pair of scales which must be kept in balance — the arms of the scales can move a little, but one arm must never go right up or right down.

This task of protecting our inner sun, our warmth, is so important that we human beings help our skin by putting extra "skins" over it, our clothes. It will help us to understand clothes if we consider certain things in nature. If for instance you watch a robin sitting on a branch on a warm summer's day, you will see that its feathers lie sleek and close against the body. But on a cold day in winter the robin sitting on a branch has its feathers puffed up.

Why does the robin puff up his feathers in cold weather? There are good and bad conductors of heat. Metal is a good conductor: if you heat one end of an iron rod, the other quickly gets hot. Wood is a bad conductor, or good insulator — I can hold a piece of wood quite safely even when the other end is burning. Fat is a good insulator. Now it may seem strange, but air is also a good insulator provided that you can somehow trap it and stop it from moving around. When the robin puffs up its feathers there is a lot of air trapped between the feathers which keeps in the warmth of the robin's body; it cannot keep it sealed off completely, but the warmth is lost slowly.

It is not the feathers, but the air between them which protects the robin's warmth. It is the same with fur where the air between the hairs protects the warmth. When we are cold, we get goose flesh. The hair on our skin rises to create a kind of fur.

However, as we human beings have little hair (and no feathers), we have always used animal-furs as a kind of extra skin to help our own skin in protecting the warmth of our body. At first people hunted animals for their skins, but later they found animals whose hair could be used without killing them; they learned to use the wool of sheep.

Sheep's wool of course also protects the warmth of the body because of the air between the fibres, but it also has another advantage; if you go out in a shower in a woollen top, the raindrops do not sink in, they stand on the surface of the coat because sheep's wool contains a fatty oil. And so you can shake the drops off a woollen top. Now think of someone caught in the rain wearing a cotton top. Every drop becomes a wet round patch — cotton absorbs the rain and then clings to the skin. It absorbs water quickly while wool absorbs water slowly. This is the reason why woollen clothes are better than cotton in cold, damp weather. The body only loses heat slowly in woollen clothes, and wool absorbs sweat slowly. Cotton, after running, feels cold and damp against the skin, and the body loses heat too quickly.

And this is a very important thing for our health: the body must not lose heat too quickly. We are bound to lose heat — the warmth continuously goes away from us into the air around — and we produce new warmth from inside. But the warmth must not go away faster than we can make new warmth. The old warmth body must not go away faster than we can make a new warmth body. If we go suddenly from a warm room to a sharp cold outside, or if on a cold day we run until we are very hot and then, feeling tired, sit down in the cold, then the body cannot make new warmth quickly enough, and we can get a chill. Some people's bodies make new warmth fairly quickly and they don't get cold often. Other bodies are slower in producing new warmth and they get cold more often.

You can also catch a cold if there are germs of a cold about — they are usually around in one way or another — and your body is unable to make new warmth fast enough. In these conditions you can catch a cold or even something much worse than a cold, pneumonia, which attacks the lungs. You will find that athletes, runners and mountaineers always put on some warm garment, like a sweat shirt or coat, after strenuous exercise that made them hot. They do this to protect the body against a too sudden loss of heat.

There is one part of the body which needs special protection against a too quick loss of heat, and that is the lower part of your body and the legs. The belly (the proper name for it is abdomen) is where food is digested. It is the warmest part of the body, and all the delicate organs which do this important task are not covered by bones, so this part is especially sensitive to cold. The legs and feet are also sensitive to cold. Our warmth body is, altogether, more sensitive in the lower than in the upper part of the body. And it is more sensitive while we are still growing.

You can mistreat the sensitive warmth body below your waist by making it use all its strength to produce warmth at great speed to make up for heat loss. But then it does not have much strength left to build your inner organs well, and when you have grown up you could be plagued by liver, kidney or stomach trouble. So it is better to keep this lower half well covered and warm in cold weather, and help these delicate parts of your body in their task.

In hot weather we must protect ourselves against too much heat outside. As we saw, dark-skinned people already have a protection in the skin, but light-skinned people must help their skins by wearing the right kind of clothes. To keep cool we need good conductors of heat, materials which let the body's heat go quickly. For warm weather cotton is a good material, and silk and linen are even better. Wool feels rough; it is like the robin's feathers puffed up. Silk, cotton, linen, feel smooth, and are like the robin's feathers kept sleek against the body in warm weather. The robin's feathers are a picture of what materials to wear in cold or in warm weather.

Artificial materials are not very good for the body. They may look nice, but they are no protection for the warmth of the body. In cold weather they let the body's heat away too quickly, and in summer they don't let the sweat evaporate.

Colour, too, has a part to play in the protection of our

warmth body. A dark coloured material lets the warmth in, while light colours reflect the heat, keeping it out. So the clothes we wear play a very important part in protecting and helping the warmth of our body, our inner sun.

11

Digestion

Without the heat of the body we could not move, we could not use our muscles, our arms and legs. This inner heat makes us independent of the outside temperature. But the warmth body is continuously dissipating and has to be renewed all the time. To renew the heat of the body we must eat. We also renew our water body and our solid body by drinking and eating. Only the air body is renewed by breathing.

So we eat and drink in order to renew the warmth, the water and the solid body. As long as we live, our body needs all kinds of building materials from outside to rebuild itself. There are no bricks in nature, so if we want to build a house, we have to take the clay from nature and change it and make it into bricks of the size and shape we want. It is the same with our food: what grows on trees and from the earth cannot be used just as it is to rebuild the body, it must be changed. An apple cannot remain an apple if it is to be any use to the body, a carrot cannot remain a carrot, an egg cannot remain an egg if the body is to make any use of it. This change which makes the right building materials for the body out of apples, carrots and eggs is called digestion.

Digestion is a wonderful process. Just think of a wolf that eats sheep all his life, but no matter how many sheep he eats, his body will remain a wolf's body and does not become like a sheep's body. And if a man were to eat sheep, mutton, all his life and nothing else, he would still have a human body. We eat fruit, vegetables, eggs, cheese, meat, and it all becomes human body, our body. And what makes all these different things which are not human into the muscles and nerves and blood of our body is the process of digestion.

Digestion takes all these things — apples, eggs, vegetables, meat — and first destroys them, breaking them up. Only when the food wed have taken in has been completely destroyed, can it be used to *rebuild* the body. And if we swallow something which the body cannot destroy and cannot break up then we cannot digest it and we suffer from indigestion.

So digestion consists of two parts: first everything we take in must be broken up and destroyed, and then it can be used to rebuild the body. The apple, the egg, the cheese, have to be destroyed until there is nothing left like an apple or an egg — but from the destroyed apple or egg our muscles and bones are built.

This work of destroying the food begins in your mouth: your teeth and your saliva already begin the destruction. When you suck a sweet, the sweet is dissolved in your saliva, and when you chew your food, the teeth grind it and break it up. What happens in your mouth is only the first part of a process that continues in your stomach and in your intestines, where all you have eaten is completely destroyed and can then be used to rebuild the body.

It may sound strange, but we start destroying some food even before we put it into our mouth: we cook it. Destroying the food we need is really quite hard work, and we make this work easier by destroying the food a little before we begin to eat it by cooking.

But what about apples, oranges, and the other food which we eat raw? We can only enjoy eating ripe apples. Unripe apples don't agree with us. What is the difference between unripe and ripe apples? The ripe apple has been "cooked" by the sun. Nearly all the food we eat is cooked: either by the sun or in the kitchen. Tomatoes, peas, carrots can be eaten raw as they have been ripened by the sun — they have been cooked for us by the sun. So that the warmth and the solid parts of the body can be renewed, so that from vegetables, eggs and meat, our muscles, bones and nerves can be built, our body must destroy food. But we start destroying the food even before we eat it, by cooking.

When we start eating, we bite off a piece of bread and

chew it — we crunch it and grind it with our teeth — which is another step in the process of destroying it. It is important that we chew each bite well; if we swallow our food only half chewed we make the work of our stomach much harder. The stomach will not be able to do its own work properly, and the whole body will suffer from bad digestion. Little babies have no teeth, they cannot crush the food before they swallow it and so they can only take in liquid or mashed food. But once we do have teeth they must be given work, otherwise they get weak. Parents give babies something hard to chew on when they get their first teeth, because they know it helps the baby to grow healthy teeth.

To keep our teeth strong, white, healthy and clean, we have to use them. Much of our food nowadays does not need a lot of work in chewing. A simple thing we can do is to chew two raw carrots every day. As they are chewed, they clean the teeth. The juice of the carrot helps the digestion in the stomach as well. So, if you care for your teeth, make a habit of chewing raw carrots. And if you want to ruin your teeth, have toothache, lose your teeth and have false teeth at an early age, then eat a lot of sweets. Natural (dark brown) sugar and honey do not harm the teeth so much. But refined white sugar is a sure destroyer of your teeth. And all kinds of confectionery and soft drinks contain this white sugar.

Teeth and Saliva

Digestion is the wonderful process by which eggs, meat or apples are changed into the muscles, bones and blood of our body. The eggs and the apples cannot change themselves into muscles or bones, nor does your stomach make muscles or bones, nor do the intestines. No, it is the spirit that lives in the body which uses the stomach, and all other parts needed for digestion, to change eggs, meat and vegetables into human blood, muscles and bones. Digestion is the work of the spirit.

Our teeth are also built up by the spirit in us — and it really is a difficult task, because the teeth are the hardest part of the body. The baby is still all soft (even the bones are still soft) when the first teeth come. The body of the little child would not yet have the strength to make hard teeth. The strength to make the first teeth is really taken from the mother: when a child is born some of the mother's strength goes into the child so that it can grow its first teeth, the milk teeth. But from the age of seven to about nine, the milk teeth are shed.

Then the body has enough strength and enough hardness to make the second set of teeth, our own proper teeth which are to last for the rest of our lives. No matter how strong and how old we get, we don't grow a third set of teeth, because later on the strength to make teeth is used for thinking! First it was strength of the body, now it becomes strength of the mind. When we grow our second teeth, it is like a sign: from now on the strength which built our teeth is no longer in the body, it is in the mind.

How is a tooth built? In the centre of the tooth there is a hollow called pulp which contains little blood vessels (that is where

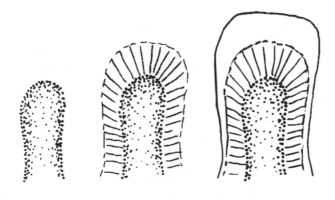

The tooth: Pulp   Pulp and dentine   Pulp, dentine and enamel

you feel pain if you have toothache). Around this hollow is an ivory-like substance with minute tubules which go like rays to the hollow; this part is called dentine. On top of and around this dentine is a very hard layer called enamel.

The teeth are also a very important part of our digestion; if they did not chew the food well, our stomach would be overworked and in time get ill. But the food is not only chewed in the mouth. Something else also happens which is just as important as chewing. When something is put on our plate which we like, our mouth "waters." Our mouth produces a liquid, saliva. Without saliva we could not swallow our food at all. Sometimes — for instance when we are very excited or afraid — our saliva dries up, and then we cannot swallow at all. But food which we like makes our mouth water, producing plenty of saliva. Food we don't like produces only little saliva and we find it hard to swallow.

Now there are all kinds of foods which are necessary but are not very tasty by themselves, and so we would not have much saliva to swallow them. And therefore we put condiments into these foods. The most important of these condiments is salt. Vegetables are made tastier with salt: potatoes, even bread, need salt. Soups need salt. In hot countries where people often don't have much appetite and only a little saliva, they use spices like pepper or curry to get the saliva flowing. And long ago the people in Europe, who ate meat that was months old, going bad and over-salted, used vast quantities of spices to make the food tastier. The wealth of Venice, the Portuguese voyages round Africa, Columbus' voyage to America, they all go back to our digestion, to the saliva which only flows if we like the taste!

But there are also other things besides condiments and spices which help to stimulate the flow of saliva in our mouth and so help our enjoyment of a meal. If the soup we are going to eat is served in a dirty plate, or if a fly has fallen into the soup, we would not feel like eating it; no saliva would come, unless you were really starving! So, if food is served on a tablecloth, with gleaming cutlery, with flowers on the table, it not only looks nice, it really helps our digestion and the flow of saliva. And if we see somebody eating noisily like a pig, it can spoil our enjoyment of the food and can spoil our digestion. We owe it to others to have table manners.

Another important thing for our digestion is that eating should be a rhythmical process, that we keep proper meal times, every day at the same time. The body wants to get a certain amount of food, and then it wants to be left in peace to get on with the job of digestion. People who are so busy that they cannot keep regular meal times often end up suffering from all kinds of stomach trouble. But if you eat between meal times, the body's work of digestion is continuously interrupted by extra bits coming down and so you have done something against your body, not for it. Your body expects a certain consideration, and if you don't give that much consideration to it you will pay for it in the end with ill-health.

Food

We need saliva to swallow our food; if the saliva dries up for any reason, we cannot swallow at all. But saliva also does something else — try the following experiment. Take a little bite of bread, don't chew it, just keep it in your mouth. Very soon the bread will dissolve and you will notice a sweet taste. Saliva is not just water, but contains a certain fluid which changes and dissolves the food. Saliva is only one of many such "digestive" juices which work in the stomach and in the intestines to change the food until there is nothing of its original substance left. Saliva is the first of the digestive juices, and if the saliva cannot work properly on the food and is not well mixed with the food, then all the other juices cannot work properly either. That is the reason why it is so important to take time with every mouthful of food and chew it well to mix it with saliva. If you gulp it down quickly, it is not chewed enough and not mixed enough with saliva, and in time this can ruin your digestion. A good many people suffer from stomach ailments simply because they gulp their food down too quickly.

When the food has been chewed and mixed with the digestive juice of the saliva, we swallow it. From the moment we swallow the food we don't really know what happens to it. Of course, we can read and learn what happens in our stomach and intestines, just as we can learn about ancient Rome. But just as we cannot go back in time and watch the murder of Julius Caesar, so we cannot watch our own digestion. We feel hungry or thirsty, we feel we have had enough food, or we feel we had too much of something. But no one can watch their own digestion at work. Everything that happens — and it is a

great deal — takes place without our being conscious or aware
of it. It is done for us, and our mind cannot command or even
watch it. We shall look at the complicated and wonderful process of digestion in the next period about the human body (see
Physiology, page 71ff). For now we shall continue to look at
what we eat.

We eat only one mineral, only one substance which does not
come from a living being, and that is salt. Water too could be
called a mineral, as it is neither a plant nor an animal, and so salt
and water are the only minerals we take in directly. Everything
else comes from the plant and animal kingdoms. There are
many other minerals which we need — calcium and phosphorus for our bones, iron for our blood, potassium, sulphur, magnesium and a whole list of trace elements — but we get all these
from the plant and animal food we eat. Salt is the only mineral
substance we take just as it is. Meat, eggs, milk or butter comes
from the animal kingdom, but the animals — chickens, sheep,
cows — have to eat plants. They build their bodies from the
food they take from the plant kingdom.

So whatever we eat (apart from salt and water) comes either
directly or indirectly from the plant world. Without plants neither human beings nor animals could exist on earth. Our whole
life depends on the plants because our food must come from a
living being, animal or plant.

But what about the plants? The plants can do what neither
human beings nor animals can do: they can take their food
from the mineral world, from the earth, from the water, from
the air (which is also a mineral). The mineral world — earth,
water, air — has no life, and only the plants can change it into
something which is alive, something which grows and has
seeds from which new plants can grow. However, the plants
can only do this with the help of the sun. Without sunlight the
plants cannot grow, they cannot change earth, water, air, into
green leaves, blossoms and fruits. It is the sunlight which works
in the plants so that they can build their bodies from dead, mineral substance.

So in the end, all that we eat — except salt and water, which
are not really food — is the work of the sunlight. You could

say that with every bite we eat, we eat some sunlight, for the sunlight is really *in* the cabbage or the apple we eat. And the warmth, the heat of your body is nothing but the sunlight that once shone on grass and cabbage and fruit.

Now we eat not only one kind of food, but different kinds of food, and each kind of food has a different purpose in the body. That is the reason why we must have a balanced diet. If you eat an egg sandwich, the bread, the butter and the egg each have a quite different purpose in your body and are used in different ways, although it all goes into the same stomach.

There are three kinds of food which the body needs. We still need other things besides the three main groups, but they come first. We need food that builds the body — the solid body — secondly, we need food that builds the warmth body, food which keeps the heat of the body going, and thirdly, we also need food that gives us strength, energy, as it is called.

There is one particular substance which builds up most of the solid body, called protein. If you want to know what pure protein looks like, you only have to look at the white of an egg, egg white is pure protein. But protein is also in meat, fish, poultry, and in milk and cheese. So protein, the substance for rebuilding the body, comes mainly from the animal kingdom. Vegetarians get protein from eggs, milk or cheese, or from plants which contain some protein, especially lentils, peas, beans.

The food which gives strength and energy comes from the plant world. The substance which gives us energy is called starch. There is starch in bananas, potatoes, rice, as well as in all grains, and therefore in flour (bread, pasta, cakes, biscuits). Sugar is also an energy food and we get sugar from plants. All ripe fruit contains some sugar, also flowers (honey), some stems (cane sugar), even roots (sugar beet). With the light of the sun plants can make sugar from air and water. When you eat starch (bread, potatoes, rice) then the starch is turned into sugar in your digestion. So the body makes its own sugar and you don't need to eat a lot of sweet things. Both starch and sugar are energy foods that come from the plant world.

The third kind of food, which we eat for our warmth, is

fat: butter, olive oil, margarine. And that can come either from animals, like butter, or from plants, like olive oil or margarine. People who live in cold climates have a greater need for fat.

Protein for building the body comes mainly from the animal kingdom. Starch and sugar, the energy foods, come from the plant world. Fat, the warmth food, comes from either. So when you eat an egg sandwich, the bread provides starch, the butter provides fat and the egg provides the protein. Each one serves a different purpose.

There is one food which contains all three — protein, sugar (not starch) and fat — and that is milk. Fresh milk also contains another substance mentioned earlier which we only take in small quantities: vitamins. Milk has all the food ingredients the body needs.

14

Bread

Everything we eat really comes from the plant kingdom; even if we eat meat or butter, the cow has built her body from the grass she has eaten.

We can eat meat or we can be a vegetarian. One cannot say, "It is bad and unhealthy to eat meat," yet there is a reason why, in one way, it is better to be a vegetarian than a meat eater. Looking at the kingdoms of nature, the mineral kingdom is farthest away from the human. It is so far away that we cannot digest minerals (except salt), only the plants can do that. Then comes the plant kingdom — it is a step nearer to us — we can eat and digest plants. The animal kingdom is nearest to us, we can eat and digest meat.

But when we eat plants, which are farther away from us than animals, it is harder work for our digestion than if we eat meat. The animal is nearer to the human being, it is less work to change meat into our own body than to change plants into our own body. Now what is better for us: the harder work of digesting plants or the easier work of digesting meat? The harder work of course, just as the hard carrot is better for our teeth than soft food. So because it is harder work, it is better and healthier to eat vegetarian food than meat. However, everyone should eat plenty of vegetables and fruit, even if they are not vegetarians.

Some plants are better for us than others. We need starch to give strength, energy to our body, and there is starch in all things made of flour — bread, cakes, biscuits — and there is starch in rice and in potatoes. The flour comes from wheat, oats, rye or rice, which are kinds of grass. Think of the way these plants grow: their slender strong stalks going straight upwards towards the sun. They don't even develop a blossom but put all their strength into the golden grain that is ripened by the sun. So when we eat bread, rice, even semolina (which comes from rice) or porridge, then we eat the grains each of which strives to become such a strong, straight plant. And the strength which might have become a golden ear of wheat or oats becomes our strength.

The potato is quite different. The part of the potato plant which grows above the ground develops blossoms and fruit, but these fruits are poisonous; we would get very ill if we tried to eat the fruit of the potato plant. But below the ground the stem of the potato plant swells, growing more and more swollen. And this swollen stem down in the dark earth, that is the potato we eat. It is not even a root, it is a tuber, a swollen stem. When we eat potatoes, we eat something that grows in darkness, something which is the swollen stem of a poisonous plant. It is therefore not as wholesome and good for the body as the golden grain ripened in the sun. Of course, we can eat potatoes, but we should not eat them too often. It is better to get the starch we need from grain.

The most important of the grain foods we eat is bread. The way bread used to be made was like this. The grain grows taking its substance from the earth, the water, the air, and the light and warmth of the sun. It is built up from the four elements which are also in us. When the corn was cut and threshed. The grain has an outer skin, the husk, which is more or less like straw, and this is removed in threshing. The grain was taken to a mill. At the mill the grain was ground between two great millstones made of solid stone, "earth." When the grain had been ground to flour, it came to the baker who mixed the flour with water — the second element — and made dough. When the dough was kneaded and ready, the baker added yeast. Yeast makes the dough rise by letting air into the dough. Then it was put into the oven, to get warmth, fire. So again all four elements were used to make bread. And bread that is made in this way really feeds and nourishes all four elements in us.

But some of the bread you get nowadays, especially white bread, is not a wholesome food. Under the outer husk of the

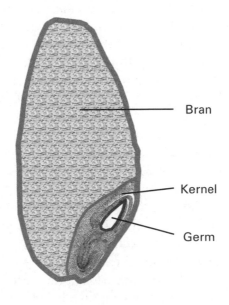

A grain of wheat

grain is a layer, the bran, containing protein and vitamin B which is needed for the rebuilding of brain and nerves. Inside the bran is the kernel containing protein and starch. And at the bottom of the grain is the germ. In this germ is the strength from which a new plant would grow. It contains protein, fat and several vitamins (B, E). Now in the old stone-mills the whole grain (without the husk which had been removed in threshing) was ground into flour. But modern mills are steel mills which do not grind the whole grain: the bran and germ are removed so that only the starch remains. All the goodness in the germ and the bran is removed (100 kg of wheat grain makes about 75 kg of white flour).

The reason for taking the best parts of the grain away is that it makes the flour white. But even when the germ and bran have been removed the flour is still not pure white. To make it pure white, benzoyl peroxide is added to the flour to bleach it. But this is not the end of the story. The old methods of making bread are too slow and unreliable: the loaves take some time to rise, and one loaf will be higher, another lower and they will not fit properly into their packaging. The supermarkets want all the loaves to look exactly the same, otherwise they will not look right on the shelves and customers will not be happy. So many other things are added to the bread while it is being made, oxidants, raising agents, emulsifiers, bulking agents, preservatives, to mention only a few. These things are not added to make the bread better for us, but better for the quick and smooth working of the machines which are used to make it.

When you buy a loaf of white bread, you not only get bread which has been robbed of protein, minerals, vitamins, of all the growing strength, but which contains all kinds of unhealthy additives. White bread not only has little nourishing value but can even undermine your health over the years. Factory-made wholemeal bread is also not really what it should be. The only way to get wholesome bread is to buy wholemeal bread from a baker or buy wholemeal stone ground flour (which uses the whole grain) and make your own bread at home. You can taste and feel the difference.

The Quality of Food

The example of white bread shows how — for the sake of better business profits — such an important food as bread is not only robbed of strength and goodness, but even has unhealthy additives put into it. And this not only happens with bread.

Human beings can only take in and digest two minerals, salt and water. We cannot eat sand or dust. If by accident some fine sand gets on a piece of bread and we eat it, the body cannot digest it. It will throw it out sooner or later. But there are some kinds of minerals which the body cannot throw out and cannot digest; these substances stay in the body. And although they are not poisonous, they are something foreign in the body, something the body cannot cope with, and so they weaken the body, making it less healthy. The body then has less strength to fight any illness that comes along. And the more of these chemical substances we get into us, the less healthy we become.

Human beings cannot change dead minerals into living substance — only the plants can do it with the help of the sun. When we make things in our factories and chemical laboratories, they are all dead mineral substances, there is no life in them. But then these factory-made things are added to the food we eat, and so they get into our body where they stay making us weaker. We do get such indigestible chemicals in our body for the following reasons.

Everything which comes from living beings, plant or animal, changes. Milk goes sour, meat goes bad, apples rot. In order to keep food for a long time preservatives and other additives are mixed in. Tinned and processed foods usually contain them. These preservatives are mostly man-made chemicals, substances which the body cannot digest and cannot get rid of. There is only a little of it in any tin of soup, beans, meat or fruit, but every tin adds a little more. Fresh food is healthier. Nature does not make everlasting food, and if we make it longer lasting we do so against nature and against the health of our bodies.

But the preservatives in tinned food are not the only chemicals. The people who make most of the food we eat know that if it has a nice colour it sells better. So butter not only gets some preservatives but also a nice light yellow dye which is a factory-made chemical. Some sausages have a nice red colour, which is a factory-made colouring.

In some kinds of ice creams, only the sugar is natural, but it is white sugar. Natural sugar is brown, and to make it white another factory process is used. There is some cream in the ice cream, but it contains preservatives. And the raspberry or vanilla taste does not always come from real raspberries or vanilla, it can also be made in a factory. And the colours are made from all sorts of chemical dyes.

There are still other ways in which man-made chemicals come into us. Free-range chickens which run about and pick their food from the ground, lay eggs with a nice yellow yolk. But battery hens are kept cooped up in cages, and hardly move at all. The eggs they lay have only a pale yolk. So battery hens get a yellow dye mixed with their food which then goes into the yolk of their eggs.

There are still other things to make us eat chemicals which are not good for us. It is not only we human beings who like fruit, there are all kinds of insects who like to eat fruit. The farmer who grows apples, pears, grapes or cherries in his orchard wants to sell the fruit and does not want to share it with insects. So he sprays the fruit with pesticide, a poison which kills the insects. This is also very harmful to birds, because insects are an important food for the birds. Many insects which do no harm, like butterflies and bees, are also killed by this poison, so in some parts of the country there are no longer any butterflies.

But a poison which kills birds and butterflies can hardly be good for the human body. It is a substance which is neither thrown out nor digested; it accumulates and over a long time this cumulative effect is harmful to us too. If you take care to wash every kind of fruit well (best with warm water) you can get rid of some of the poison, although many pesticides enter the flesh of the fruit and cannot be removed.

This example shows you that our modern methods are not only a danger to human health, but interfere with nature, wiping out useful insects and birds together with the harmful ones. But the worst interference with nature, and one which is most frightening, is done by some farmers in the way they treat the soil.

Plants draw the most important food they need to grow from the earth on which they grow. Now in a forest the leaves fall down in autumn, they turn into earth again, and so all that was taken from the earth is given back to it. But with a field of grain everything that has grown from the soil is taken away when the grain is cut. And after two or three years the soil has no longer anything on which plants could feed. The soil becomes lifeless and you cannot grow anything on it. Farmers have always known this. And to give back to the soil what it needs, they made compost heaps of all the vegetable material that wasn't used (potato peels, kitchen refuse) as well as cow dung. There is nothing in the whole wide world that contains more good things for the soil than cow dung. And that compost heap or manure heap — this mixture of cow dung and all kinds of plant material — after a time turns into rich, good earth. Everything in the compost heap came from living things, from animals and plants.

But then some clever people said: "In our chemical factories we can make something which will not only make your soil as it was before — it will make it better; the fields that used to give 10 bags of corn will yield 20 bags of corn if you use our stuff." This "stuff" made from chemicals in factories is called chemical fertilizer.

Chemical fertilizers did work. Farmers were delighted, for every 10 bags they had before, they now got 20 bags or even more. But two other things happened. First, the nourishment value of this grain became less and less. There were less vitamins and proteins in wheat and in every vegetable grown with

these fertilizers. Secondly, the soil was not really improved by
the chemical fertilizers. It became poorer and poorer and after
several years not even tons of fertilizers could make anything
good grow from it. The soil had become so dead that it was
ruined.

Today many farmers are turning away from these fertilizers,
and are using compost and manure again. These organic farms
produce a much better quality of food, and you can taste the
difference between a normal (chemically-fertilized) cabbage or
carrot and an organic one. You will also find much more wild-
life — birds, butterflies, flowers — living around such farms.
But the chemical fertilizers ruin the earth — and they cheat our
body.

Physiology

Dr Harvey and
Blood Circulation

If you had lived in the sixteenth century, as a child you could have seen Columbus and Leonardo, and later you could have seen Queen Elizabeth I, Mary Queen of Scots, Shakespeare, Francis Drake, Luther, Calvin and John Knox.

The next century (1600–1700) was not so blessed with people of genius. To meet Cromwell or the Merry Monarch, Charles II, or even Peter the Great of Russia would not have been as interesting as meeting Leonardo or Francis Drake. The seventeenth century was much more a time of turmoil, of upheavals and civil wars, of people of the same nation fighting each other. In such times of bloodshed and fighting, there was little interest in art or in knowledge. You cannot expect great works of art in Germany when the whole country was a battle-field in the Thirty Years' War. You cannot expect astronomers to calmly watch the stars while Roundheads and Cavaliers fought their battles. And yet science *did* make some great strides in the seventeenth century in spite of the battles and wars.

While King James VI (or I of England) wrote books about the divine rights of kings, Copernicus, a priest far away in a part of Germany that is now in Poland, wrote a little book which said that the earth was going round the sun. While William of Orange was fighting the Spaniards, a Dutchman, Zacharias Janssen, discovered that by looking through two lenses, a fara-way church steeple could be seen as clearly as if it were only a few yards away. This invention of the telescope was very useful to William of Orange in his fight against the Spaniards. But in

Italy a lonely scientist, Galileo Galilei, used the telescope to look at the stars. He was the first man to see the moons of Jupiter and the phases of Venus. So in all these troubled times the quiet work of scientists and thinkers went on, and their work was perhaps more important than the battles and wars.

There was another great man of science who lived at that time. William Harvey was a doctor in England at the time of Charles I, the king who was later executed by order of Cromwell. William Harvey was a very good doctor, who had studied medicine not only in England but also in Italy which at that time had the best teachers in the science of medicine. William Harvey had become so good that Charles I made Harvey his personal physician. It was a great honour in those days to be doctor to the king.

Since his student days Harvey had been interested in the way the heart and the blood work in the human body. The doctors and scientists of that time did not know as much of the heart and the blood as we know today, and some of the ideas they had were quite wrong. In those days what doctors knew about the human body came not from observation but from ancient books. Far back in Roman times, when Rome was a great empire, a doctor called Galen had written a book about the human body. And fourteen hundred years later, at the time of Harvey, doctors all over Europe still considered the teachings of Galen as a truth that could not be doubted.

According to these ancient books of Galen the blood moved in the body with an ebb and a tide, just as the sea has a low tide and a high tide. Only they follow each other much quicker of course. At rising tide the blood supposedly flowed from the heart to all parts of the body. Then when the blood reached your fingers, your toes or your ears — according to Galen — it evaporated, as water evaporates. But then a new tide would come from the heart bringing new blood to all parts of the body. That was what Galen taught, and at the time of Harvey, doctors still believed in these tides of the blood.

Leonardo da Vinci was one of the first men to doubt this theory of the blood's tides. He cut up corpses to find out about the mysteries of the human body, and from his study of the blood vessels, he did not think that this old theory could be right. But Leonardo was not a doctor, and the doctors took no notice of the ideas of a man who was only a painter.

Almost two hundred years after Leonardo, Harvey too began to doubt this ancient theory of the tides of the blood. He said to himself: every pulse beat, every heartbeat is one so-called tide. The human heart can hold about 85 ml (3 fl oz) of blood — it is about the size of a fist — and in that space there is only room for 85 ml of liquid. With every beat the heart is supposed to send 85 ml of blood to all parts of the body. In one minute the heart beats 72 times; 72 times 85 ml is about 6 litres, and in one hour that would be 360 litres (about 800 lb) of blood, which is several times the whole weight of the body. The theory could not be true: the body just cannot make new blood at this rate of 360 litres every hour. The only possible answer was that the same blood which goes out from the heart also returns to the heart, that the blood circulates.

It is like the circulation of water in the world. The water comes down from the clouds as rain, flows down mountains and hills to the sea and rises again to the clouds. The water which comes from the clouds is the same water that was once in the river and sea. And the blood in the body also circulates, going from the heart and returning to the heart.

Harvey not only thought about it, but studied the bodies of frogs and fish as well as the bodies of criminals who had been hanged. He studied and observed for years until he was quite certain that his idea of the circulation of the blood was true. Then he wrote a book about it, but when this book was published, all the doctors turned against Harvey. Who was he to doubt the wisdom of Galen and who was he to think he was more clever than the great Galen?

The entire medical profession turned against Harvey calling him mad. Of course, Dr Harvey was very unhappy about this and he was also worried, for if King Charles would also think he was mad, he might dismiss him and Harvey would lose his job as royal physician. How could he show the king that he was not mad and that the blood did circulate?

Now one person's misfortune can be another person's good

luck, and so it was with Harvey. At that time a young man lived in England who some years earlier had fought a duel. In this duel he had received a terrible wound in his chest, but he was a strong young man and recovered from the wound. However, he was left with a hole the size of a fist in his chest. It did not worry him, for it was always covered with bandages and he lived quite happily with this hole in his chest.

Dr Harvey knew this young man and invited him to his surgery. He also invited King Charles, and when the king arrived, Harvey took the bandages off the young man's chest and asked the king to look into the hole. The king not only saw the heart beating — contracting and expanding — but also saw tubes which pulsated. He saw the blood vessels which lead blood away from the heart (the arteries) and those which lead the blood back to the heart (the veins). And King Charles said to Dr Harvey: "I know now that you are right, and in time the other doctors will see that too."

Dr Harvey remained a royalist all his life. He was very sad when Charles lost the civil war and later his life. But by then there were already some doctors who had come round to share his idea of the circulation of the blood. And when William Harvey died at an old age, he knew that before long all doctors would know he had been right. His name will always be remembered as one of the great scientists of the world: the man who discovered the circulation of the blood.

The Threefold Human Being

The solid matter of which the table is made, remains the same as long as the table lasts, but the stuff of which our body is made, bones and flesh and blood, is slowly but continuously changing. In about seven years there is hardly a single particle in our body which is the same. So after seven years our body is not the same as it was, but each one of us is still the same person. What makes us the same person is the soul. No matter how the body changes, there is still the same soul. So although I have changed the substance of my body about eight times in my life, I am the same person I was forty or fifty years ago.

But where in the body is the soul? The soul is not visible — but neither is the magnetic force in a magnet, for an ordinary piece of iron and a magnetic piece of iron look the same. To find out if there is magnetism I can put iron filings near the pieces of iron. One of them will attract the iron filings. That is how we know that there is a magnetic force in one of the pieces of iron. And to find out how and where the soul lives in the body, we have to observe what the soul does. Our soul does many things even without our knowing about them. But we will start with things we do know.

Imagine you are in a garden and see a flower and you recognize this flower as a rose. And to recognize the flower as a rose you have to *think*; you have to remember what other roses look like. It is not a difficult thought — not like algebra — but it is thinking just the same. But you are not only thinking and recognizing the flower as a rose, you may also enjoy looking at its beauty. You *like* the beauty of the rose. Liking or disliking is quite different from thinking: it is a feeling. And now you can break off the rose, take it home and put it into a vase. That is neither thinking nor feeling. That is doing, for which we must engage the will. In this simple event — I see a flower and recognize it as a rose; I like its beauty; I pick the rose — we have the three main activities of the soul: it thinks, feels and wills.

As we go through the day there is not a single moment when we are not doing all three of these things. There is always some thought in our mind, we are always feeling something — we feel interested or bored, liking or disliking — and we are always doing something using our will. As long as we are awake, we are always thinking, feeling and willing. When we go to sleep, we cannot think, but we can feel, and these feelings are our dreams; dreams are feelings. We shall see later that we can and do use the will in our sleep.

We have thinking, feeling and will, and we have waking, dreaming and sleeping. And we can see how these correspond.

But what we really want to know is how this is connected to the body. We know that thinking has to do with the head, with the brain. If somebody gets hit hard on his head he faints, he loses consciousness; he stops thinking. The brain is the instrument of thinking. The brain does not think by itself — it is like a violin or a piano, which cannot play by itself. It needs a musician to produce music. But the musician also needs the instrument — he cannot make music without the violin or the piano. The soul is the musician who needs the brain to produce the "music" which we call thinking.

The brain is not something by itself — from the brain millions of nerves go to all parts of the body; without these nerves the brain could not work at all. The brain and all the nerves together are called the nervous system. (When we say that a person is "nervous" we mean that they are anxious or worried, but this is not what is meant by "nervous system". It just means the system which is made up of nerves.) The whole nervous system is the instrument for thinking. The brain is the centre of this nervous system, and the nervous system goes over the whole body.

Now just as the nervous system is the instrument of thinking there is another system which is the soul's instrument for feeling. To find out what this system for feeling might be, let us think of a very strong feeling, for instance a feeling of shame which makes us blush. Some people blush very easily — even when there is nothing to be ashamed of — they have very sensitive feelings. Or take another very strong feeling: a sudden fright, when the opposite of blushing happens and we become pale, we blanch. But what does blushing or getting pale mean? It means that either more blood than usual comes to our face (blushing) or less blood than usual (blanching). From this we can see that the blood is the soul's instrument for feeling.

It is not only the blood, but also the breathing which takes part in our feelings. Notice how our breathing changes, getting quicker when we are excited, or the sudden gasp when we are frightened. The blood and the breathing, both are the soul's instruments for feeling. Our blood circulation and our breathing together, are the second system, the system of feeling. Now as we breathe in a rhythm and as the blood moves through the body in a rhythm (you can feel it in your pulse beat, that is the rhythm of the blood), this is called the rhythmic system. The rhythmic system, the soul's instrument for feeling, reaches to every part of the body, but it has its centre in the heart and in the lungs, just as the nervous system had its centre in the brain.

When we were in the garden and picked the flower, we walked to it with our legs and had to use our arms and hands to pluck it. So it is clear that the limbs have something to do with the will. If we do some hard work or some sport that we are not used to, we find the next day our arms or legs or some other part of the body are sore. What is really sore are our muscles. This shows where the will is at home. It is in our muscles. And we not only have muscles in our arms and legs but all over the body and even inside your body.

We can take a piece of bread in our hand and put it into our mouth, we chew it and swallow it. Up to this moment we can decide what our muscles should do, but now other muscles inside the body take over, which work on their own. The gullet pushes the bread down into the stomach (even if you stand on your head the muscles of the gullet would still push the bread "upwards" into the stomach). In the stomach muscles work on the bread, and then there are muscles in the intestines or bowels.

All these inner muscles are far more important than the muscles in arms and legs. For if these inner muscles would not work, if they would not help to digest the food, then our arms and legs would have no strength at all and could do nothing. So all the muscles of our body depend on the muscles of our stomach and intestines, on the muscles which work for our digestion. Without the digestion of food, no muscle in the body could work, just as the nerves could not work without the brain. And so the whole system of muscles is called the digestive system. It is the third system, the soul's instrument for the will or action.

So we see how the soul lives in the body. It uses the nervous system for thinking, the rhythmic system for feeling and the digestive system for the will.

Thinking	Feeling	Will
Nervous	*Rhythmic*	*Digestive systems*
Waking	*Dreaming*	*Sleeping*

We must remember that although they are separate systems, each goes over the whole body. They work together, side by side, in every part of your body. In our little finger there are nerves, there are blood vessels and there are muscles: all three systems are there. In ancient Greece, Plato, a pupil of Socrates, said the soul was like a charioteer driving three horses. The three horses are the three systems of the body.

The Three Cavities
of the Body

The body consists of hard parts like bones, and soft parts — the muscles and the inner organs like lungs, heart, liver, stomach and so on. But the hardest parts, the bones, are arranged to form cavities or caves in which the soft parts are enclosed. There are three main cavities. The first is inside the skull; it is the cavity for the brain. The second cavity is inside the chest, the ribs form the cavity for the heart and lungs. And the third cavity is inside the belly and it contains the stomach, liver, intestines. The brain, in the first cavity, is the centre of the nervous system. Heart and lungs in the second cavity are the centre of the rhythmic system. And the stomach and intestines are the centre of the digestive system. Even the cavities of the body are arranged to house the centres of the three systems.

There are differences between these three cavities. The first, the skull containing the brain, really is a cave, for the brain is surrounded and protected on all sides by the strong bones of the skull, so that nothing at all from outside can get to the brain. And the bones which form this cave, the skull, are the hardest bones of the body.

But the second cave is not a real cave at all, because the ribs forming this cavity are a little apart from each other. They are like the bars of a curved cage and are named the ribcage. So the heart and lungs, the centres of the rhythmic system are in a cage rather than in a cave.

And the third cavity is not formed by bones but is only enclosed by strong skins with only some bones at the back, the lower bones of the spine. The stomach, liver, intestines, are only separated by strong skins from the world outside. They are not surrounded by rock-like bones as the brain is, they are not even in a cage like the heart and the lungs, but are in a bag.

This is a very wise arrangement, for it is necessary that the cave of the brain is different from the cave of the stomach. The stomach and the intestines are muscles as we have seen, and muscles move when they work. The stomach and the intestines move a great deal when they digest the food. We know nothing of it, but the muscles of the stomach and intestines *do* move, and work hard in the process of digestion. The brain, on the other hand, does not move at all. No matter how hard we think, our brain remains still. In fact, the brain simply could not move, for it has no muscles and is completely enclosed by the skull.

So there is the cave at the top, the highest cavity of the body, and the brain lies quite still and motionless in the skull. Lower down, in the region of the belly, is another kind of cavity formed only by strong skins and muscles. And inside this cavity of the belly the stomach and intestines work and move vigorously. Between them there lies something partly closed in by rib bones and partly by skin: the ribcage. Inside the ribcage are the heart and the lungs which do not lie still like the brain, but move regularly according to certain rules — they move *rhythmically*.

The three cavities are different: a cave, a cage and a bag. And what happens in the three cavities is also different:

— there is no movement of the brain in its cave,

— there is rhythmical movement of heart and lungs in their cage,

— there is strong movement of stomach and intestines in their bag.

And each of the cavities is the centre of a system which goes through the whole body. The motionless brain is the centre of the nervous system, the beating heart and lungs are the centre of the rhythmic system, and the moving stomach and intestines are the centre of the digestive system.

All three are instruments of the soul. The brain and whole

nervous system are the instruments of thinking, and it is strange that the instrument of thinking is so completely surrounded by hard bones. By contrast the stomach and intestines, the instruments of will, are only separated from the world outside by a strong skin and muscles.

You can understand this if you realize the difference between thinking and will. There is a difference between thinking of an apple and holding a real apple in your hand. When you think of the apple, you are quite independent of the world around you, rising above the real world; your mind can form a picture of an apple whether there are any about or not. The instrument for thinking also keeps aloof from the world, so the brain keeps a solid wall of bone between itself and the world. But when you do something, when you handle an apple, you must be in touch with it; you must be right down to earth for any action of will. And so the instrument of will, the stomach and intestines, are not aloof or cut off from the world by bones, but just separated by skin and muscle.

Thinking and will are opposites. Thinking rises above the world, while the will goes right into the world. You can even see it with people. There are people who are very fond of studying and thinking who can be very clever, but are not very practical and cannot do anything useful with their hands. And there are practical people who are very good at making things with their hands, but have no mind for study and learning. The first live too much in thinking, the second too much in the will.

So it is not surprising that the brain is sheltered on all sides by bones which keep the world out, just as the thinker withdraws into his study and does not want to be disturbed. But the stomach and intestines, which have to handle and work on the real food we eat, are not so withdrawn. They keep much closer to the real world and are only protected by a strong skin. And in the middle, between the opposites, there is the rhythmic system, the instrument of feeling. In feeling we are never aloof from the world nor are we really doing or changing anything. And so the cavity for the centre of the rhythmic system is neither a cave nor a bag, but a cage — something which is not quite closed.

Just as there are people who live very much in their thinking and people who live very much in their will, so there are also people who live very much in their feelings. You often find such people among artists. A painter for instance has to go out and look at things, and then comes back to his studio to paint. He can make sketches outside, but the painting is done inside: he lives in a rhythm.

In ancient India people were divided into three castes. There were the priests, the holy men, whose real task was thinking. They did not just pray, but did the thinking for all the other people who came to them for advice in all things. These holy men did not do any work, they lived in a temple or in a cave, just as the brain lives in a cave. The second caste were not artists but warriors whose task was to defend the land or make new conquests. The warriors did not have to be wise but had to be brave, to have courage, which is a feeling. The warriors did not live in temples or caves but in castles from which they rode out to battle and to which they returned. And the third caste were the peasants or farmers who tilled the land and harvested the crops. A farmer did not have to be wise or brave, but had to be hard-working with strong muscles. They lived in huts which had only thin walls.

In those ancient times in their social life people followed the rules of the human body and soul.

Thinking	brain	cave	priest	temple
Feeling	heart	cage	warrior	castle
Will	stomach	bag	farmer	hut

In our time we must not be one-sided, but be all three at the same time: wise, brave and hard working; we must use our brains, our hearts and our hands equally well. Just as the three cavities make one body, so wisdom, courage and strength should be one in us.

Digestion and the Stomach

The three systems of the body — the nervous, the rhythmic and the digestive system — are the instruments of the soul, just as the violin is the instrument for a musician. But the soul has three instruments and plays all three at the same time — it plays on the nervous system for thinking, on the rhythmic system for feeling and on the digestive system for will.

But the violin, piano or clarinet are very simple instruments compared with the wonderful instrument of the human body and its three systems. The instruments made of wood and metal cannot be compared with the living nerves and blood and muscles. We shall now look in more detail how these instruments of the soul work, beginning with the digestive system.

Our will always involves some kind of movement. When we walk we move our body, when we write we move a pencil or pen, when we build a house we move bricks, when we speak we move the air. To make any of these movements we use our muscles. What happens if we use our muscles to run? We get warm. And the same happens whenever we use our muscles strongly. If we feel cold we can move our arms quickly to get warm. So we can see that moving the muscles produces warmth, heat. On a cold day we have sometimes shivered. This shivering is nothing but little muscle movements to produce heat and raise the temperature of our body.

But every muscle movement, even if we just move a finger or turn our head, produces a little heat, a little warmth. There is no muscle movement of any kind without warmth, without heat. Without this heat the soul cannot use the will in the body. On a cold day it can happen that our fingers get stiff with cold; our soul cannot move the fingers, for there is not enough warmth. We must first warm our hands before the soul can reach into the fingers again.

The soul is not something physical like a stone or water or even air, and it is only through heat, through warmth, that the soul can use its will in the muscles. Heat and will belong together. Some people have very strong will-forces: they can easily get into a boiling rage, or are hot-tempered, as we say. Such people show us that heat and will belong together. And a man who is "hot with anger" will not sit quietly, he will bang the table or shake his fists and will move his muscles in one way or another.

We can see now how the will of the soul gets into the muscles: by heat, by warmth. But heat does not come from nowhere. We have to burn something, and that is what really happens with every muscle movement. Something is burned in the muscles to produce the heat. Even if we only move a finger, a little combustion takes place in the muscles of our finger to produce the heat without which the finger could not be moved. And what is burned continuously in all the muscles of our body comes into the muscles from the food we eat. We eat and digest the food so that our muscles have something to burn.

Now we can see why the whole system is called the digestive system. Digestion gives all the muscles the food and the heat they need. One could just as well call it the heat-system of the body, because what is really needed for the will is the heat, but there would be no heat without eating and digestion.

When and where does digestion begin? You may think it is in your stomach or in the mouth? No, it begins with the cooking! When we cook food, we change it, preparing it for the body. We heat it which means we begin the burning of the food outside the body. But we don't burn it completely, just partly to make it easier for the body.

What happens when we eat raw fruit? The fruit must be ripe, that is to say it has been cooked by the sun: it is the heat of the sun that has "cooked" the apple or the orange for us. But whether it is an apple or a piece of bread, it has been slightly burned before our body takes it in to continue the burning.

After the first step in digesting the food in the sun or the kitchen has been done, we come to the next step. First we chew the food with our teeth. We have three different kinds of teeth to do this work efficiently. The front teeth show by their shape that their task is to cut. (When we bite an apple we use our front teeth.) There are the eye-teeth or canines at the sides which have the task of tearing the food into smaller bits. Then at the back we have molars which grind the food like millstones grind flour. The front teeth simply take a bite; they don't do the heavy work, so they are like the head! The eye-teeth only divide the bite into smaller bits, and one could compare them with the rhythmic system. But the really hard work of crushing the food until it is ready to be swallowed is done by the back teeth, the molars which are most akin to the digestive system. So even our teeth are threefold, as the whole human being is.

As we saw earlier (Chapter 13), the crushing of the food is not the only thing that takes place in the mouth. When something is put on our plate which we specially like, our mouth "waters" — plenty of saliva comes from special glands in the mouth. Without saliva we could not swallow our food at all. Sometimes, if a person is very excited, his saliva dries up and he cannot swallow at all. So the saliva is necessary to make our food go down. But the saliva also contains a digestive juice; it changes the food. That is why a piece of bread, if you leave it in the mouth, begins to taste sweet. The digestive juice turns starch into sugar. And only when the food is made soft by the teeth and thoroughly impregnated with saliva is it ready to go on the next part of its journey.

And now comes that phase of digestion of which we are not conscious at all. Even the swallowing is a far more complicated process than we might think. At the back of the mouth there is an opening, but from this one opening, two tubes lead downwards. One, the gullet is for the food. The other, the windpipe, is for breathing. We also use the outgoing air of the windpipe for speaking.

Both the windpipe and the gullet have only one opening at the back of the mouth. At the very moment we swallow food the windpipe closes itself, so that no food goes down the wrong

way. If we eat too hastily, or talk with our mouth full, sometimes something goes down the wrong way, into the wrong
pipe. Then the lungs quickly send up air, we cough and splutter,
and the bit of food is thrown out of the windpipe.

The closing or covering of the windpipe takes place every
time we swallow, although we are not conscious of it. From this
moment onwards muscles start to do their work and we don't
know about it. The gullet is not simply like a rubber tube, down
which the food falls, but all along the gullet there are muscles
which squeeze the food down, as we squeeze toothpaste from
the bottom to the top of a tube. That is why we can swallow
even when standing on our head. The food is squeezed to the
stomach; it does not merely slip down.

The food comes into the stomach. The walls of the stomach
contain strong muscles that contract and expand rhythmically.
They tighten around the food, pushing, squeezing and rolling it
from side to side: they work on it with great vigour. And at the
same time they mix the food with strong digestive juices. These
gastric juices come from glands in the stomach and are strongly
acidic. Through the work of the muscles and the gastric juices
the food is changed, becoming more and more liquid until it is
like thick soup.

Then the food is ready for the next stage and goes into the
intestines. The intestines again add more digestive juices and
have muscles which squeeze and push the food, until it is quite
liquid. Only now is it ready to be sent into the body.

21

Digestion and the Intestines

If you slip and fall, before you have time to think, your hands come forward and they take the worst of the fall to protect your head. Your hands and arms don't wait until you want to move them — they move on their own by your unconscious will. It is always the will which moves the muscles. The will can be conscious, as when you lift something or work or walk, or it can be unconscious, as when you fall and your hands shoot forward to take the worst of it.

Now we can do many clever things with our conscious will, but the unconscious will is really much wiser, because all the work done by the stomach and by the intestines is done by the unconscious will. The stomach and the intestines contain muscles, but these move through the unconscious will. This unconscious will can even be lazy, for instance when the bowels don't work properly and you get constipation. Or this unconscious will can overwork and then you get the opposite of constipation, diarrhoea. So this unconscious will that works in the digestion can be lazy or overworking, making you ill. But this unconscious will does not over or underwork without a reason.

If you eat too hastily without chewing your food properly, without letting the saliva work on it long enough, then you make the work for the stomach harder. If that part of the digestion is not already done in the mouth, then the stomach becomes like someone who is made to do more than his fair share of work — it gets into a bad mood. A good-tempered stomach contracts and expands in a beautiful rhythm, but a bad-tempered stomach contracts and expands wildly and not in a smooth rhythm. Some people often over-eat, then the stomach becomes overtired and hardly works at all, contracting and expanding only a little. This unconscious will, that moves the muscles of the stomach, does not get lazy or ill-tempered without reason.

But if it is not upset by the silly things we might do, then this unconscious will works with a wisdom that is far greater than the cleverness we have in our heads. The food going from the stomach to the intestines is turned into a liquid. The intestines are really a kind of kitchen where the unconscious will in us cooks and prepares the food for the body. Our home kitchen is really the first stage of the cooking that goes on inside the intestines. We cook food first to make the inside cooking in the intestines easier.

Just as a cook puts all kinds of seasoning into the food, the invisible cook in the intestines also needs all kinds of seasoning, all kinds of ingredients. These extra ingredients needed in the intestines are sent there from other parts of the body. For instance the gall-bladder, which is beneath the liver, produces a terribly bitter juice, called bile. The intestines need this bitter bile for their cooking for it dissolves fat. Without this bitter juice we could not digest any fat. If we have trouble with the gall-bladder and it stops sending bile to the intestines, we cannot eat any fat. It would just make us sick. So the bile from the gall-bladder is very necessary for the cooking that goes on in the intestines.

Another organ that sends many special ingredients to the kitchen in the intestines is called the pancreas. It sends a digestive juice which is an alkali. In the stomach there is a strong acid (hydrochloric acid), but the juice from the pancreas is an alkali containing many different ingredients which can digest carbohydrates, fats and proteins. The bile from the liver dissolves fats, but it is the juice from the pancreas which digests it. So the different ingredients all work together. (The pancreas also has another job: it makes insulin. Without it we would get an illness called diabetes, in which we cannot properly control our body sugar). The cooking that goes on in the intestines is quite complicated; it is far more complicated than the cooking in our home kitchen. And the cook who does all this is the unconscious will in us.

When all this cooking has been done, when the food we have eaten is really ready, it has become liquid. No matter what we eat, in the intestines it becomes a liquid, a kind of thin soup. But the intestines are only the kitchen for the body. Now this food must go from the kitchen to all parts of the body, to all the muscles which need the food to make heat. The muscles cannot move without heat, and they cannot make heat without food. So the food has to go out of the intestines to all parts of the body. How the food, once it is ready, goes from the intestines to all parts of the body is arranged quite wonderfully.

The intestines are one long hollow tube, an elastic tube that lies in coils in the cavity of our belly. If this long, elastic tube were taken out and uncoiled and stretched out in a straight line it would be about 6 metres (20 ft) long, but this great length is nicely coiled up inside us. Surrounding this long tube there are thousands and thousands of tiny blood vessels, and the tiny blood vessels lead to larger vessels which eventually lead to all parts of the body. So the food which is now a liquid has only to get from inside the tube to the outside, to these tiny blood vessels where the blood takes to all parts of the body.

How does this happen? The inside of the tube, the inside of the intestine looks and feels like velvet. Velvet is made up of tiny hairs of wool, but the lining of the intestines is not hair or wool: it is made up of thousands and thousands of tiny little tongues. They are tongues which taste the food-liquid, and only when these tiny tongues *like* the taste of the food-liquid is it allowed to leave the intestine and be taken by the blood to all parts of the body. These tiny tongues are called villi (singular: villus).

So the unconscious will in you not only cooks the food in the intestines but, like a good chef, first tastes the food before serving it. These tiny tongues, the villi, are the tongues by which the food is tasted. For instance, the skin of apples or grapes or the stringy fibres in cabbage does not nourish the body, and the villi, the tiny tongues, will not let this cellulose through. It stays inside the intestines which push this stuff further and further, until it is pushed out of the body altogether when you go to the toilet. The reason we go to the toilet is to get rid of the matter the villi have rejected. However, this fibre, as it is called, is necessary for without it the intestines would become lazy and weak.

But the nourishing matter in the food-liquid which the villi like, is allowed to go out of the intestines. But how the food goes out of the intestines is not a simple thing. There are no holes, no pores in the intestines, not the tiniest ones. Yet, the strange thing is that slowly and gradually the food-liquid disappears from inside the intestine, and little drops of it appear outside in the blood vessels. Some food is absorbed and some is changed in this process, but there are complicated processes which scientists even today do not yet fully understand.

The little blood vessels outside the intestine receive the food-liquid, and the blood carries the food from the tiny blood vessels to a large vein, the portal vein, which carries the food-liquid first to the liver. The liver is still another "taster" checking if there is too much sugar in the blood. If the blood is too sweet, the liver takes the extra sugar out and stores it. The liver is a store for sugar not needed at the moment. But when you run, the muscles burn a lot of sugar, which they take from the blood. This lowers the sugar-level in the blood, so the liver releases sugar from its store into the blood.

It is strange to think that just below the liver which guards the sweet sugar, lies the gall-bladder which makes the bitter bile for the intestines.

All this cooking, tasting and seasoning is done by the unconscious will working in the intestine, in the liver, in the gall-bladder and in the pancreas, so that the whole body can live and move.

Breathing

From the moment we chew our food until the blood stream brings the food to all parts of the body, to all the muscles where it is burnt to produce heat, this whole journey is about 24 hours. So our whole digestion keeps time with the sun.

Perhaps this is not so surprising. Nearly all the food we eat comes from the plant-world. Even if we eat meat, butter or eggs, the cows or the chickens have fed on plants, and have built their bodies, their milk or eggs from the plants they ate. Plants grow through the light and warmth of the sun. And when we digest our food and burn it in our muscles, we release the heat of the sun which once shone upon the plant. One could say the plants capture the heat of the sun, and we set it free again when the food is digested and burnt in our body. So there is a connection between digestion and the sun, and our digestion takes 24 hours on average (sugar takes less, fat more time).

In this heat of the body our will lives. When we run, we use more effort and more will than when we stand or sit, but when we run we are also hotter than when we stand. The will-forces of our soul really live in the heat of the body. The body always has a certain heat, a temperature of 37°C (98.6°F). One could say there is always a gentle, invisible flame burning in our body which never gets so hot that it burns the body and never so cool that it leaves the body stiff and cold. In this gentle, constant, invisible flame there lives the will.

This constant heat of the body is quite a mystery. If you burn coal in a fire it is nearly impossible to keep it at an even temperature for hours — it gets hotter or cooler all the time. But the burning of food in our muscles keeps an even temperature all the time. Furthermore, two thirds our body is water, so this burning is going on as if underwater. This is not quite as impossible as it seems. In chemistry we can pour two liquids together in a test tube and the tube becomes quite hot. Something similar but much more complicated takes place in the body to produce the gentle constant heat in which our soul lives with its will.

Although it is a complicated chemical process which burns the food in the body, it is a burning. Whenever something burns it needs oxygen, and the burning which takes place in our muscles, although it is quite different from the flame of a candle, also needs oxygen. The food could not be burnt without oxygen. And that is why we breathe.

We breathe in oxygen so that the food in all parts of the body can be burnt. Just as when a piece of wood is burnt and partly becomes carbon dioxide, so when the food in the muscles in all parts of the body is burnt, part of it becomes carbon dioxide, which we breathe out.

However, there is a difference between the burning of a candle and the burning that goes on in our body. The burning candle takes oxygen from the air and sends carbon dioxide back into the air at the same time. But the human body does it in a rhythm, first taking in oxygen, then sending out carbon dioxide. In-breathing and out-breathing are done in a rhythm by all living creatures. The candle is dead and without life or rhythm, but living creatures take in oxygen and send out carbon dioxide in a rhythm.

We now come to the rhythmic system of the body. We saw how the three systems — nervous, rhythmic, digestive — work together. If the rhythmic system would not provide oxygen and take away carbon dioxide, the digestive system could not work at all.

The rhythmic system is the system of feeling. And we can see this in the breathing. If we feel very upset and cry, we sob. This sobbing is very short *in*-breathing. When our feelings are upset, our breathing too is upset and instead of the steady rhythm, there is a short, broken rhythm of sobbing. But when we laugh, which is the opposite kind of feeling, then we breathe *out* more strongly than normally. So again a strong feeling breaks the steady rhythm of breathing. Or another example: if someone feels sad about something they sigh — they breathe deeply. All these and other examples show you that our feelings work into the rhythm of breathing. But even feelings which are not very strong work on the breathing, changing it a little. When a lesson is more interesting, you breathe a little quicker than when it is not very interesting.

But when we run, breathing also gets quicker; why? Because then our muscles are burning more food, they need more oxygen, and the breathing hurries to give the body the extra oxygen. On the other hand, when we are asleep our breathing slows down because the muscles rest and we only need oxygen to keep the body at its even temperature. The normal, average time of breathing is 18 times per minute, though during the day we often breathe quicker, and at night we breathe slower.

On average we breathe 18 times per minute, which is $18 \times 60 \times 24$ breaths every day, or about 25 920 breaths per day. The earth also breathes: the plants coming out in summer are like out-breathing, and their disappearance in winter is like in-breathing. The plant world is the lungs of the earth. But the great earth breathes slowly: it takes a whole year for one complete breath. And as we know from astronomy, when the great earth has made 25 920 breaths, then the spring equinox, the point where the sun is on March 21, has gone through all twelve constellations of the zodiac returning to where it was before. Our human breathing, our human rhythm, is a little miniature of the great cosmic rhythm between earth and sun.

This breathing-rhythm which is so timed that it is a miniature of the cosmic rhythm, the Platonic or great cosmic year, works back upon the body into the rhythm of our heartbeat and blood stream. If you count the pulse beats in one minute, you will find that there are about 72 pulse beats, that is 4×18. To every one breath there are four pulse beats, just as there are four seasons in each year. If you run and your breathing gets faster, your pulse beat also gets faster, so that there are always four pulse beats to one breath. The rhythmic system — breathing and blood circulation — is full of wonders.

Breathing and the Blood

Blood circulation and breathing are linked by the rhythm of four to one: four pulse beats in the blood to one full breath. The human being needs rhythm, just as there are the great rhythms of the cosmos — day and night, summer and winter, and the great cosmic year of 25 920 years. The rhythms of our own body are linked to the rhythms of the cosmos — our breathing is a tiny miniature of the slow, great breathing of the earth. And the pulse beat is linked to our breathing rhythm, just as the four seasons to one year.

Because there is rhythm in our body we respond to and enjoy the rhythm in music, in poetry and in dance. Not only the mind but the whole body enjoys rhythm in music and poetry, and the body becomes stronger and healthier by singing and reciting poetry. Of the three systems — the nervous, rhythmic and digestive — the rhythmic system is the great health giver, and anything you do rhythmically helps and strengthens our health.

Breathing and blood circulation are not only linked by this rhythm, but also by the work they do for the body. We shall follow the journey of the air from the moment we breathe in to the moment we breathe out. Very many things happen in that short time of less than two seconds between in-breathing and out-breathing.

The journey of the air begins in the nose. We can of course also breathe in through the mouth — and when we are short of breath we do — but there are good reasons why we should breathe in through the nose. Our body takes nothing in from outside without first changing it. We have seen all the things that happen to food before it is passed to all parts of the body. Similarly air is not taken in just as it is. The air around us is nearly always colder than the temperature of the body, and it would not be good for the lungs if they had cold air coming in all the time. The lungs would suffer and get ill if that happened. So there are very many tiny blood vessels inside the nose, and the blood in these vessels warms the air as it goes past them.

As you can imagine, its needs a lot of blood to warm every breath, and there is a lot of blood in these tiny vessels. They must be very near the surface to do their job. If you bump your nose, you bleed much more than if you cut your finger, as these tiny blood vessels easily burst. If you climb a high mountain where the air is thinner, where it has less pressure than in the valley, your blood pressure is still as strong as it was down below and it can happen that a blood vessel in the nose bursts and starts bleeding. The reason that there is such a lot of blood in the nose is to warm the air breathed in.

The nose not only warms the air, it also acts like a filter to remove the dust which is always in the air. The nose has two lines of defence against the dust. There are tiny hairs inside your nostrils which catch the larger dust particles. And then inside your nose there is a very sensitive skin. You only have to tickle this skin inside your nose with a feather to find out how sensitive it is. And when this sensitive skin is even slightly irritated it produces a sticky fluid called mucus. And the tiny particles of dust get stuck in this sticky fluid and do not get to the lungs with the air. If you have a cold, this sensitive skin is irritated all the time — it is inflamed and goes on producing that sticky mucus whether it is needed or not.

When you cry, tears come from glands inside your lids. Normally the glands produce only a little liquid to wash the eyes free of dust, but when you cry, you produce a lot of this tear liquid. Some of it goes through little tubes into your nose where it irritates the sensitive skin which produces mucus, and then you have to blow your nose.

Once the air has been warmed and the worst of the dust has been caught by the little hairs and the sticky liquid in the nostrils, the warmer and cleaner air goes from the nose into the windpipe which leads to the lungs. When the air we breathe in comes down the windpipe it is still not clean enough for the lungs. There is another cleaning device inside the windpipe. Growing from the walls of the windpipe are fine threads called cilia that are never still, continually sweeping up and down. They catch any fine dust that is still in the air, and when they have a little collection, we feel a tickle in our throat and cough and so get rid of this dust. That is why we start coughing hard if we inhale a cloud of dust.

As we have followed the journey of the air we have discovered why we sneeze and cough. The windpipe goes down from the neck into the chest. Halfway down the chest it divides into two branches, just as a tree trunk divides. But this particular tree trunk, the windpipe, is hollow and it is like a tree growing downwards. The windpipe divides into two branches, one goes to the right, the other to the left lung, for we have two lungs, one on each side of the chest.

In the lungs each branch divides into smaller branches, and the smaller branches divide into still smaller ones — it really is like an upside-down tree. These branches are called bronchi which simply mean branches in Latin. Bronchitis is an illness in which these tiny and tiniest branches are inflamed.

A tree has fruit at the end of its branches, and this upside-down tree also has a kind of fruit at the end of its tiniest bronchi. It has thousands and thousands of tiny hollow globes called air-cells or alveoli. These thousands and thousands of air-cells together are the lungs. The windpipe is the tree, the bronchi are the branches, and the lungs are like thousands of fruit on the tree. The natural and healthy colour of the lungs is pink, but the lungs of smokers are covered with soot and tar and turn black.

We have followed the journey of the air from the nose to the windpipe, and through the bronchi and into the air-cells. They are called air-cells because they get filled with air every time we breathe in. But that is only one half of the journey of the air, for in each little air-cell there are tiny blood vessels, and the blood in these blood vessels takes the air from the air-cells. And now the bloodstream, the blood circulation takes over and carries the air to all parts of the body and to all the muscles, which need it to burn the food. And the bloodstream takes the used air, the carbon dioxide, from all parts of the body and carries it back to the air-cells in the lungs and now the journey goes up the bronchi and the windpipe and we breathe out.

Oxygen and Carbon Dioxide

We breathe in oxygen and we breathe out carbon dioxide. Oxygen is the fire-giver, for without oxygen a flame dies and there is no fire and no warmth. For human beings oxygen is therefore the life-giver, for without oxygen we would die in a few minutes. Carbon dioxide on the other hand is a deadly gas for people as well as for animals. In Italy a cave called the Dogs' Cave, has a layer of carbon dioxide rising from the volcanic ground. Because carbon dioxide is heavier than air, it rises only to a certain height, about the height of your hips. If people walk into the cave they are in no danger because their heads are above the carbon dioxide and can breathe life-giving oxygen, but if a dog is taken into the cave it dies quickly because its head is in the carbon dioxide layer, where it suffocates.

Life-giving oxygen is what we breathe in. It goes down the windpipe, through the two branches, into the little bronchi and into the little alveoli or air-cells. At one moment these tiny balloons are filled with oxygen. And in the next moment the oxygen is gone, having been absorbed by the blood. But as the oxygen is taken away, carbon dioxide takes its place. The blood has brought the carbon dioxide to the air-cells, so all the time the blood is taking oxygen from the air-cells and putting carbon dioxide in its place.

It is the blood which brings oxygen from the alveoli or air-cells to all parts of the body, and it is the blood which takes carbon dioxide away from all parts of the body. We are only aware that we breathe in and out, but the real work of bringing oxygen to all parts of the body and removing carbon dioxide, is done by the circulation of the blood. It is the circulation of the blood which keeps us alive.

If human beings and animals kept on taking oxygen from the air and giving back carbon dioxide, there would soon be no oxygen left and all human and animal life would die out. But this does not happen, because the plants absorb carbon dioxide and produce oxygen. Their breathing is just the opposite of ours.

The plants have not got lungs like we have. Their "lungs" are different: we have all seen the lungs of the plants for they are the leaves. Plants breathe with their leaves. A leafless tree in winter has stopped breathing, it is hibernating like some animals do (a hibernating animal hardly breathes). The tree starts breathing again in spring when the young leaves come out. But if a tree is stripped of all its leaves in summer — as occasionally happens in a great gale — it will die — for the tree must breathe at some time of the year.

How do the leaves breathe? There is a special substance in the leaves which takes the carbon dioxide from the air and breaks it up, dividing it into carbon and oxygen. We all have seen this wonderful substance — it is what makes all plants green. It is called chlorophyll.

It is this wonderful green substance which breaks up carbon dioxide into carbon and oxygen. But it can only do it with the help of sunlight. Chlorophyll is such a special substance because the light and the warmth of the sun can enter right into it. The light of the sun, which has come into the chlorophyll in the plants, divides carbon dioxide into carbon and oxygen. At night, when there is no light, chlorophyll does not work and the plants do not give out oxygen.

During the day the sun's light and warmth enter into the chlorophyll and together break carbon dioxide up into carbon and oxygen. Oxygen goes out into the air while carbon remains with the plants. Throughout the day more and more carbon is gathered in the plants: the plant grows. The plant grows by absorbing more and more carbon from the breaking up of carbon dioxide. When we eat food, we eat the carbon of which the plant is built, and when the carbon (strictly the carbohydrate, which is starch or sugar) is burnt in our body, we release the sun's light and warmth that once entered the green chlorophyll.

So chlorophyll and sunlight do two things for us: they give us the oxygen we breathe and the carbon we eat.

Only the plants have this substance into which sunlight can enter. Human beings have no chlorophyll. If we had, we could change the carbon dioxide in our own body by ourselves. But then we would be green like the plants and sunlight would enter into us as it does into the plants. We are not green as we have no chlorophyll in us, but we have something like chlorophyll, called haemoglobin. There is one great difference between haemoglobin and chlorophyll. Chlorophyll contains a little magnesium, while haemoglobin contains a little iron which makes the haemoglobin red. The redness of the blood comes from nothing but the red haemoglobin in it. Chlorophyll makes the plants green, while haemoglobin with its iron makes our blood red.

It is the red haemoglobin in the blood which takes the oxygen from the alveoli or air-cells and brings it to the food, the carbon in the muscles. It is haemoglobin which brings carbon and oxygen together so that they become carbon dioxide, and through this heat is released. In the green chlorophyll of the plants the carbon dioxide is again divided into carbon and oxygen and the sun's heat is captured. The red blood and the green plants are opposites or polarities. They belong together, complementing each other as night and day complement each other. In chlorophyll the sun's light descends, in the red blood the sun's energy rises again. To understand the human body you need the rhythms of the cosmos and the strange breathing of the plants. The human being needs the whole world.

Blood Circulation, Liver and Kidneys

A tree outside takes in carbon dioxide from and gives out oxygen to the air which surrounds it. The upside-down tree through which we breathe — windpipe, bronchi and air-cells — takes in carbon dioxide from and gives out oxygen to the blood, to the body which surrounds it. This tree behaves like a real tree outside, though of course it has no chlorophyll, and it does not transform the carbon dioxide; only the real plants can do that.

The blood takes oxygen from the lungs and also returns carbon dioxide to the lungs. But the blood cannot go from the lungs with oxygen and return with carbon dioxide by the same route: it must do it on two different routes. So it goes from the lungs on one route carrying oxygen to all parts of the body where it picks up carbon dioxide and returns by another route to the lungs. On both routes there is a kind of station through which the blood passes. Whichever way the blood goes it must pass through the same station, and that is the heart. Coming from the lungs with oxygen, the blood passes first to the heart and then goes out to all parts of the body; and returning from all parts of the body with carbon dioxide it first passes through the heart before going on to the lungs.

The routes on which the blood moves on this journey are of course tubes: thick tubes and thin tubes and tubes so fine that they can only be seen under a microscope. Each thicker tube branches into smaller and smaller ones like the bronchi so that the blood can really reach into every part of the body. All these tubes are called blood vessels. The tiniest blood vessels, those which only show under a microscope, are called capillaries which mean "little hairs," though they are even finer than hairs.

The capillaries are really the "turning points" of the blood. In the capillaries of the lungs, the blood gives up carbon dioxide and starts on its next journey with oxygen. In the capillaries of all other parts of the body, the blood gives up oxygen and starts on its return journey with carbon dioxide. The blood vessels which go to and from the heart are the largest ones — as thick as a finger.

With all these thicker and thinner blood vessels and capillaries, there are two different kinds: those taking oxygen from the lungs, and those returning carbon dioxide to the lungs. That is quite easy to understand. However, the doctors who first discovered this circulation and gave names to the two kinds of blood vessels, did not start from the lungs, but from the heart. The blood vessels, which carry blood away from the heart, are called arteries and those which lead back to the heart are called veins. This can be muddling, as the vessel bringing oxygen-rich blood to the heart from the lungs is a vein. Yet all other veins carry blood with carbon dioxide. It is the same with the arteries which all go from the heart carrying oxygen, except the one artery from the heart to the lungs which carries carbon dioxide.

The blood, when coming from the lungs with oxygen is quite different from the blood returning with carbon dioxide. The oxygen-blood in the arteries is bright red, but the blood returning in the veins with carbon dioxide is more bluish-red. The arteries with the bright red blood are elastic, stretching with every beat of the heart. This gives the pulse which you can feel in an artery near the skin. The arteries have muscles to widen or narrow, letting more or less blood through. But the veins in which the bluish blood flows no longer have any noticeable pulse, and have far fewer muscles. The bright red blood in the arteries is like a young man going full of vigour and courage out into the world. But when the bluish blood returns through the veins, the young man has become old.

The whole circulation of the blood is like a miniature reflection of the four seasons. When the blood flows away from the lungs bright red with oxygen it is summer, when it returns, laden with carbon dioxide and bluish, then it is winter. And in the capillaries of the lungs — where the carbon dioxide is left behind and new life, new oxygen is picked up — it is spring, while in the capillaries of the body, where the blood gives up oxygen and absorbs carbon dioxide, it is autumn.

We must not forget how food is taken by the blood from the intestines to all parts of the body. The blue blood in the veins, which takes carbon dioxide from the body also takes the food-liquid from the intestine. It takes the food-liquid to the liver which works on it and changes it, then to the heart, and from the heart to the lungs. In the lungs the food-liquid goes with the blood which is now refreshed with oxygen, first to the heart and then to all parts of the body.

So the blue blood, when it takes food from the intestines, brings it first to the liver. The liver is the guardian who takes away any surplus sugar. The red blood, before it brings the food to all parts of the body, takes it first to another guardian, the kidneys. The kidneys have a very fine sense of taste. Unfailingly they taste if there is something in the blood which might be use-less or harmful for the body, and they send this out through the bladder. For instance, when you digest the protein in your food, some waste called urea is produced in the liver. The blood car-ries this dissolved urea to the kidneys and the kidneys send it to the bladder for disposal in the urine.

The kidneys are also affected by how you feel. When you are very excited about something they make more urine than usual, and you have to empty your bladder more often.

The Healing Power of Blood

The oxygen flows from the heart in the arteries, and the carbon dioxide flows to the heart in the veins. But it is the same blood which flows in both. It changes from one stream into the other in the tiny little capillaries. In every part of the body — in the little toe and in the tip of the nose — there are arteries, veins and capillaries. If you look carefully in a mirror you can even see them in the corners of your eyes. Try to have a picture in your mind of these surging streams: the bright red stream coming from the lungs into the heart and going out from the heart, splitting into smaller and smaller streams reaching every part of your body. But the tiniest streams of one river reach into the tiniest streams of the second river; as the blood passes into the other capillaries it turns from bright red to bluish red. And the tiny streams come together in larger and larger river-beds in which the blood flows back to the heart and from the heart to the lungs. But this blue river also carries the food, first to the liver and then back to the lungs, and the red river carries the food-liquid, first to the kidneys and then to all parts of the body.

The red stream distributes the food we have eaten to all parts of the body. But now we come to the real wonder of this food distribution. Different parts of the body need different kinds of food: the muscles need protein to build them up and they need carbon (in the form of sugar or glucose) to burn; but the bones need calcium to grow strong and hard; the brain needs phosphorus. There is hardly a part of the body that does not need something special out of all the food we eat. While scientists have only discovered what each organ needs in the last two hundred years, the blood has known it all the time.

There is a wisdom in the blood that knows exactly which substance and how much of it any organ needs, and it delivers just what is needed. The blood is not just a fluid flowing to and fro, but has a wisdom which knows that the bones need calcium and the muscles need protein. Just imagine if it were to deliver calcium to the stomach, the stomach would become stiff and hard like a bone, but the wise flow of the blood makes no such mistakes. (Though calcium is sometimes deposited in the arteries, and contributes to arteriosclerosis, hardening of the arteries, a common disorder in older people.)

And the same wise stream of blood is also the great healer of the body, doing more for the body than any physician can do. Just take something very simple: you cut your finger. If the blood did nothing about it and just flowed on, then even the tiniest cut, a pinprick, would make you bleed and bleed, and eventually you would bleed to death. But the blood does something about it. Normally the blood is a thin liquid like red water. But when you have cut yourself, the blood around the wound — and only around the wound — changes. First it becomes sticky, then gradually the sticky substance thickens and hardens becoming a crust. This crust seals and closes the little blood vessels which the cut has opened, and so the bleeding is stopped.

This thickening and hardening of the blood around the wound is called clotting, and if the blood did not have this wisdom to clot when necessary, we would die of the tiniest pinprick. There are people whose blood does not clot (they are called haemophiliacs) and they must carry an artificial clotting substance with them at all times.

But this is not all. When you cut yourself, the skin has been slit open. The blood makes the edges of the wound grow towards each other until they meet in the middle and the wound is closed. When the cut is too deep and wide for the edges to grow together, the blood first forms new blood vessels in the gap which help to form a new skin to fill the gap between the edges. With small wounds the cut disappears completely, but with deep wounds the new skin is a little different from the original skin leaving a scar. So the wise blood not only clots to stop bleeding but also repairs damage and grows new skin. When a bone is broken, it is again the blood which heals the bone with new bone-substance.

Take another example. Sometimes a foreign substance, for instance a little splinter, gets deep under the skin. The blood first surrounds the invader with a special fluid which slowly presses the splinter to the surface of the skin. The skin swells with the pressure (that's why it hurts), and then breaks throwing out the invader. This special fluid is pus, and is just another of the healing manifestations of blood.

Sometimes the body is invaded not by a little splinter, but by a whole army of enemies: the germs (either bacteria or viruses) of some infectious disease like measles, mumps or diphtheria. These germs poison the body. Whenever such a poison invades the body, the blood makes a counter-poison, something which counteracts the first. These wonderful counter-poisons — we might call them medicines — are called antibodies. The amazing thing about these antibodies is that they are different for each invader. Mumps is a different poison from measles, and the blood makes just the right medicine, the right antibody for each invader. Once such an antibody has been made there is always a remnant of it in the blood and it fights any second invasion from the start. That is why people very rarely get measles or mumps twice: after the first invasion the blood remembers which antibody to use.

Almost all healing in the body is done by the blood. Whatever the illness, sore or disease, it is the wise river of the blood which cures and heals. The human doctors only learn from the blood, and all medicines and pills are meant to help the blood in its great task. But the real wisdom of healing is in the blood.

The blood is continuously renewed. Through all its wonderful work the blood becomes old, tired and dies, and new blood is formed. In about four months the entire blood is renewed. The place where the new blood is formed is inside the bones. If you look at a chicken or pork bone, you see that it is hollow. In the hollow of the bones is a soft substance called marrow, which creates new blood, so that in about four months the entire blood is changed and renewed.

One could say the blood not only heals the body, it heals itself, discarding what is old and tired and renewing itself from the marrow of the bones. Blood is truly the great healer.

The Heart

Blood, the great healer, is intimately linked with our breathing. Breathing and circulation are part of our rhythmic system. And in our whole rhythmic system lives our feeling. Our feelings change the flow of the blood. Shame or embarrassment drives more blood to our face, but when we are afraid, less blood comes to our face and we become pale. Sometimes a person has a great shock and faints. Why did they faint? In such a sudden case, so little blood comes to the head that not only the face but the brain does not get enough blood. If the brain does not get enough oxygen and nourishment from the blood, it stops working and we faint becoming unconscious.

So our feelings work on the flow of the blood: strong feelings work strongly, though every feeling works to a greater or lesser degree on the blood. Even very slight feelings work upon the heart. A worry for instance is not a very violent or strong feeling. Some people who have great responsibilities worry a lot, and of course there are also people who worry about small things. But if this goes on for years, these worries weaken the heart. Their doctors warn them to stop worrying and avoid any excitement, or the heart will simply stop. All the urgency and rush of our time put a strain on the heart, as do bad feelings like envy, hatred, greed. But if you work with real love and enthusiasm, then you strengthen the heart.

The heart feels what is going on in our soul. Some of our moods are like warm sunshine to the heart while others are like a cold blast. The heart not only feels what is going on in the soul, it also feels what is going on in the body. If we eat and drink too much, the heart feels it and becomes lazy and weak. If we are restless and do not relax from time to time, the heart feels it and becomes stiff and hard. The heart feels how the stomach works and how the liver works; and just as you would share the joys and sorrows of a friend, so the heart shares the joys and sorrows of every organ in your body. The heart is the great friend of every organ in the body.

How does the heart know what is going on in the stomach or in the liver or in the skin? The heart cannot leave its place, but it learns what is going on everywhere in the body from the blood. The heart "feels" the slightest change in the blood and every little change is a message of what is going on in the body. For instance, after a meal the stomach and intestines work harder and need more blood; the heart feels this and sends more blood to the lower part of the body and less blood to the head. That is why we cannot do complicated mental work after a heavy meal: the brain cannot get the extra blood, for it is needed in the stomach and intestines. It is the heart which regulates the blood supply to every organ.

The blood flows twice through the heart on its journey, once going to the body and once coming from the body. As the blood goes through the heart, the heart feels it and learns what there is to learn about how all the organs are doing and what they need. The heart is the centre of the blood circulation and it is also the centre of our feelings.

Something happens in the heart which happens nowhere else in the body: the flow of the blood is stopped, for a brief moment between every heartbeat. Between one heartbeat and the next there is a little pause, and in that pause the blood is stopped and the heart can feel what it must know to keep the body healthy. During the course of a minute all the blood in the body passes through the heart. Every minute every drop of blood in you has been stopped in the heart and has given its message.

How is this stopping and then going on of the blood is done in the heart? The heart is hollow of course, so that the blood can flow in and out. It is divided into four parts or four chambers. The partitions or walls between the four chambers form a kind of cross. But only the partition that goes down the middle is a real wall. It is a very thick strong muscle, which divides the heart into a left half and a right half. The left half is for the red blood that comes from the lungs and passing through the heart goes to all parts of the body. The right half is for the blue blood that comes from the body and goes through the heart to the lungs. The blue and red streams never mix in the heart, but are kept apart by the strong wall of muscle in the middle.

The vertical partition in the heart divides the red blood from the blue blood, but the horizontal partitions are not walls, they are valves which open and close, only allowing blood to flow in one direction. The two top chambers, called atria, receive blood and then contract. Their pressure sends the blood through the valves into the ventricles (the lower chambers). The top parts, the atria, expand again to receive new blood, and at the same time the ventricles contract, their pressure sending the blood away. Then the ventricles expand to be ready for the new blood which the contraction of the atria will bring.

So the heartbeat comes about in this way: the top part, the atria contract and at the same time the bottom part, the ventricles, relax and expand. Then the atria expand and at the same time the ventricles contract. If you listen to the heartbeat through a stethoscope, you can hear the beat: "lubb-dupp — lubb-dupp — lubb-dupp." The "lubb" is the sudden closing of valves caused by the contraction of the strong ventricles. The softer "dupp" is the closing of valves caused by the contraction of the atria. And the pause is the short fraction of a second when the heart "feels" the blood and so gets to know what is going on in the whole body.

As we saw, the heart feels in two ways: it feels what is going on in the soul — excitement makes the heart beat faster and worries weaken the heart — and it also feels what is going on in the body, for instance after a big meal sending more blood to the stomach and intestines. But the heart feels all this through the blood. The blood contains the food we have eaten or drunk, and some of these things are felt by the heart very strongly. Coffee, for instance, works on the heart like a whip, making it beat faster. Of course, in the long run, it is not good for the heart to be driven harder. Tea has a similar, but not quite so strong, effect on the heart. Alcohol and tobacco also have their effects.

The heart feels what is in the blood, and so knows what is going on throughout the whole body. But to know anything, we need a nervous system and brain. The heart also has a little nervous system and brain of its own: it has a little knot of nerves in the muscles of the upper part on the right side. This little knot of nerves is called the pacemaker, for it regulates the heartbeat. The heart is the only organ in the body that has a little brain of its own.

But now you can recognize what a complete organism the heart is. It is, first of all, a muscle, and muscles belong to the digestive system. The contracting and expanding of the heart muscle, the beating of the heart, are no different to the work of the muscles in the arms or legs, or from the movements of the stomach; they are all muscle movements. But unlike other muscles, the heart muscles go on unceasingly as long as we live. The heart has muscles, it has a little nervous system of its own, and of course the blood in the heart belongs to the rhythmic system. The heart is a complete being: with a digestive, rhythmic and nervous system. It is as if nature had placed a complete little being inside the whole human being.

In most popular books about the human body the heart is called a pump because the expanding and contracting of the heart do indeed help to move the blood, to pump it through the body. But if you think of the heart just as a pump, then you forget that the heart also stops the blood for a moment, you forget that the heart has a nervous system or little brain of its own, and you forget that the heart feels and shares our excitement, our joys and our worries.

What we must keep in mind is that the heart is the centre of the whole blood circulation, the centre of this wise circulation that flows ceaselessly through veins, arteries and capillaries. Imagine an optical instrument which would show only the flow of the wise circulation of the blood. You would see the complete human shape, for through the capillaries the blood goes into every part of the body, but everything would be in continual movement, from the lungs and heart to the periphery and back again.

The Nervous System

Imagine a similar instrument which shows not the blood, but only the nervous system. The nerves are fine threads or fibres which have their centre in the brain and reach from the brain to all parts of the body. We would again have a picture of a complete human shape, but in this shape there would be no movement at all. In the digestive system there is the movement of the muscles, in the rhythmic system there is the rhythmic flow of the blood, but the nervous system has no movement of its own. And there is a reason for this stillness.

In a fast flowing mountain stream you cannot see a reflection of the trees or of the sky, but in a quiet lake or pond where there is not even any wind to ripple the water, you can see mirrored a faithful image of sky and trees. Our whole nervous system is still and quiet, because its task is to be a faithful mirror of the world. If your nerves and your brain were as full of life and movement as your blood, you would know as little about the world around you, as you know about what's going on in your blood. The whole nervous system is a kind of quiet, unmoving mirror which gives us a faithful picture of the world around us.

The nervous system is the soul's instrument for thinking, as well as sensing: seeing, hearing, tasting, smelling and touching. The first task of the nerves is to have a true picture of the world through our senses. Only then can we think about it. All our sense organs — eyes, ears, the sense of touch in our skin — belong to the nervous system.

But we must be quite clear about one thing. It is not our eye which sees the world, it is our soul which sees the world through the eyes. There was an interesting experiment made

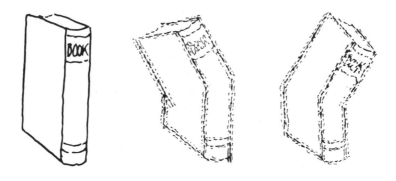

Left: Book seen normally. Centre: As seen through distorting glasses.
Right: Book as it appears after removing distorting glasses.

at the University of Innsbruck in Austria which showed that it is the soul or mind which sees the world and not only the eye and the nerves. Some volunteers, mostly students, were given special spectacles to wear for several weeks. They were not ordinary glasses, but distorting glasses which broke straight lines and made rainbow colours along the edges.

At first the people who wore these distorting spectacles were really lost; they could not walk alone, they could not lift a pencil, for it was not where they saw it. They were quite helpless. After three or four days they said things did not look quite so crooked any more and they were no longer quite so helpless. After six weeks — although they still wore the spectacles — they saw everything just as we see it: straight, and even the rainbow colours had disappeared. And they could walk, eat and do things just like everybody else.

When they were able to see quite normally, the spectacles were taken off and they saw a crooked world, but crooked in the opposite direction, and they saw rainbow colours along the edges (but the colours in opposite order). The eyes had adjusted to correct the wrong picture given by the spectacles. But after a few days things looked a bit more normal, and after a few weeks they saw things as they really are: they had regained "normal vision."

It is the soul which teaches the eyes how to see and what to see. It is the same with the eyes of a baby whose eyes are just as good as an older person's. During the first weeks of a baby's life its hands cannot get hold of anything it wants to grasp. It takes time before the soul has trained the eyes to see properly. And the same thing happened with people who had been born blind but whose eyesight could be restored by an operation later in life. They did not see the world as we see it; they saw only floating blurred shapes, and only weeks later, did they see the world as we all see it. The whole nervous system — the eyes, the senses, the nerves — is only a wonderful instrument of the soul which is prepared by the soul to function properly.

The Eye

We looked at our sense of sight and how the eye is trained by the soul to see in the right way. Under the eyelids there is the cornea, a tough transparent skin, a domed window which protects the eyeball. Under the cornea is the grey, blue, black or brown eye, called the iris (the Greek word for rainbow). In the centre of this coloured circle, the iris, is a dark hole, the pupil. In strong light the pupil becomes smaller so that not too much light gets through. In dim light the pupil is enlarged to catch as much light as possible.

Behind the pupil is the lens. It is like a glass lens, but is soft and changes shape. Looking at distant things it becomes narrow and thin, while looking at near things it becomes rounder and thick. The light going through the lens produces an upside-down picture of the world on the back of the eyeball. It does not matter that the picture is upside down. Our soul puts it right, as it corrected the distorted pictures people saw in that experiment. The inside of the eyeball is filled with a transparent jelly which keeps the eyeball round. If the eyeball were hollow, it might lose its round shape, distorting and blurring the picture at the back.

At the back of the eyeball, where the picture appears, there is the retina, a special kind of skin that takes coloured pictures, like a film in a camera. But the retina takes a picture continuously. The retina contains some special chemical substances which are sensitive to light. For example, there is a beautiful rose-red substance called visual purple (or rhodopsin). This chemical is very sensitive to light and we need it for our night vision, our ability to see in very poor light. There is another, similar (but not so sensitive) substance we need in order to see in daylight, called visual violet (or iodopsin).

If you look at a very bright light you cannot see properly — a spot or a black circle floats in front of your eyes. When bright light falls on that visual violet (iodopsin), it becomes clear. The bright light clears away the visual violet, and you see nothing where the light came in. And this "nothing" is the black hole, the black circle you see floating before your eyes. You could never see again if the blood did not heal and restore the visual violet. There are fine capillaries reaching into the retina and blood, the great healer, restores the visual violet. So after a little time, the black spots in front of your eyes disappear and you can see normally again.

A little of this destruction of the visual violet happens whenever you look at something. Only when you look at less bright things, the destruction of the visual violet is very slight and is so quickly repaired by the blood that you don't even notice it. But without this continual destruction and renewal of the visual violet you would not see at all. Everything you look at makes a hole in the visual violet and this hole is repaired by the blood in about $^1/_{15}$ of a second. This is the secret of how the retina can take picture after picture. Each picture destroys the visual violet, the blood restores it and the old picture is wiped out.

There are certain occasions when we can notice something of this healing and repairing work done by the blood in the eye. When you look at a red circle and then at a plain background you see a green circle, the after-image. This is also a kind of healing: the opposite colour is produced when there is too much of one colour. The blood works on the retina, which was damaged by the red, and you notice the effect as the green after-image.

Our eye is not like a dead camera, but is active, creating something like these after images. Sometimes our eye is far more active than we know. If we draw a line and divide it exactly in the middle, our eye tells us that the dot is really in the middle. If we add arrows, they make the eye more active and the one looks smaller than the other. It is an optical illusion. There are many examples of optical illusions which all show how active the eye really is.

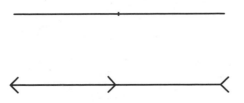

A line divided in half, but with added arrows one half appears larger.

We look at the world through two eyes and it makes quite a difference that we have two eyes. If we look at something close by first through one eye, then the other, we notice that the two views are not quite the same. When we look with both eyes at the same time, the two different views come together and things look "real" to us. They are three-dimensional and not flat like a painting. This binocular vision is important to us as it enables us to judge distances correctly. If you hold the pencil in your left hand, close one eye and quickly try to touch the pencil with a finger, you may need several attempts, because one eye cannot judge distances.

Here is another experiment of binocular vision. Roll some paper into a tube and hold the tube before the left eye. Keep both eyes open. Now hold the edge of your right hand to the outside of the tube. It seems as if you had a hole in your right hand. This is, of course, because the two views, left and right eye-views are merged.

Another important function of the eyes is so-called "persistent" vision. If you look at something, and then quickly look away, the first picture does not disappear immediately. The retina keeps the picture for about $1/15$ of a second after you look away. We can see this from following experiment.

Fold a piece of paper in half. On one half draw a head with the mouth open, on the other half trace the same head but with the mouth closed. Roll the top half round a pencil. Now use the pencil to roll the top half quickly backwards and forwards over the bottom half.

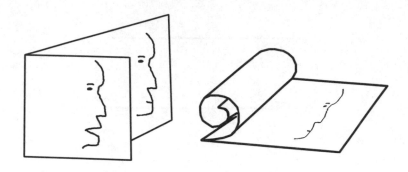

Creating the illusion of movement from two still pictures.

You see only one face, opening and closing its mouth. Films and TV work on the same principle. They are still pictures moved so fast before our eyes that the old picture is still there when the new one comes, and we "see" normal movement. Our retina keeps the old picture for about $\frac{1}{15}$ of a second. It sounds a very short time, but without it there could be no films, they would flicker unbearably.

Another interesting fact about your eye is the so-called blind spot. This blind spot has to do with the nerve which goes from the retina to the brain. Take a piece of paper, make a dot and about 10 cm (4 in) to the left make a cross. Close your left eye, focus on the cross. Move the paper a little closer or farther away and at the right distance the dot suddenly disappears. It lies in the blind spot of the eye.

All the wonderful activity of the eye would be of no use to us, and we would not see anything, if there were not the optic nerve going from the retina to the brain. If by accident this nerve is damaged, we would be blind even if every part of the eye itself is in perfect order.

The left eye and the right eye each have their own nerve to the brain, but it is complicated. Part of the nerve of the left eye goes to the right side of the brain and part of it goes to the left side. The nerve from the right eye goes partly to the left side and partly to the right side of the brain. The two optic nerves partly cross.

We might think that the spot in the retina where the optic nerve begins is a spot where we will see particularly well, but it is not. Where the nerve begins, we don't see at all; that is the blind spot we discovered. That is one of the strange things about the eye: we cannot see without the optic nerve, but just where eye and optic nerve come together we are blind.

But without this optic nerve we could not see. And it is so with all our senses. Our ears could not hear without nerves going from inside the ear to the brain, our nose could not smell without nerves going from inside the nose to the brain, and our fingers would have no feeling for rough or smooth without nerves going from the fingers to the brain. What does a nerve look like? It looks like a kind of whitish fibre, though not hollow like a blood vessel. Our sense organs, eyes, ears, are only the outside frontiers of the nervous system.

When you look into the blinding light, a dark spot came dancing before your eyes; not everything was dark, only this spot. This dark spot does not mean that you see something; quite the contrary — where the dark spot is you see nothing. At least temporarily your eye does not work. While the eye is not working, the blood repairs the damage. That part of the eye is really asleep for a moment, and during this time the blood can heal and repair the damage.

Everything you see does a little damage, some slight destruction to the visual violet, to the retina and even to the nerve that goes on from the retina to the brain. If there was not this slight destruction going on in the eye, in the nerve, we would not see at all. It is the same with all our senses and with all the nerves: we would not see, hear, smell, touch or taste if there was not a slight destruction going on in the senses and in all the nerves. And the blood keeps on repairing and healing the damage so quickly that normally we never notice what is going on.

But the blood cannot completely repair all the damage which is done. All the time sights, sounds, smells and touch are coming in, and a little damage in the nerves and in the brain remains unrepaired. When all this damage has reached a certain point then the whole nervous system does what your eye did when it looked into the bright light, it stops working and we fall asleep. And while we are asleep, the blood does not have to cope with new sights and sounds and can repair all the damage it could not cope with during the day. Sleep is like the black spot you see after looking at the light, but it encompasses the whole nervous system.

Now we can understand something very important. The rhythmic system does not need any sleep. The heart and the blood go on working day and night. The digestive system does not need sleep. The stomach and intestines go on working day and night. The stomach even complains if it has no work to do, and then we feel hungry. Only the nervous system needs sleep. It needs sleep to repair the slight destruction which goes on all the time we are awake. However, without this process of destruction we could not be awake at all. To be awake is really a kind of illness, a very mild illness which is cured every night in our sleep. And if people don't get enough sleep they get really ill. To be awake is a kind of illness, but without this illness we would know nothing about the world and would be blind and deaf. But every night we become temporarily blind and deaf when we sleep and the illness is cured and the damage repaired.

The Ear

The outside part of your ear is designed to catch as much sound as possible; people who don't hear well cup their hands to their ears to increase the "sound-catcher." And even the thicker and thinner parts of the ear are so designed that they catch the sound and bounce it into the earhole. Then the sound travels through a tube, the auditory canal, until it reaches a thin skin called the eardrum that is really like the tight skin on a drum. The eardrum is not beaten with drumsticks, but it is drummed on by the sound which the air brings from outside.

The sound now needs a kind of amplifier. This is done by three tiny little bones which together look rather like a limb. One tiny bone that is like the hand, touches the eardrum — vibrating with the eardrum — and the other two bones that are like the arm magnify the movements of the vibrations. In this way, the vibrations are amplified. These amplifying bones should really be called, hand, arm and shoulder, but are actually called hammer, anvil and stirrup.

The third bone — the shoulder or stirrup — touches another tight skin called the oval window. You can see that through the three bones, the vibrations of the eardrum now come drumming on the second skin, the oval window. The oval window is the window of a strange house shaped like a snail's shell, the cochlea, which is Greek for snail's shell.

The ear is made up of three parts, the outer ear, the inner ear, and between them the middle ear. The outer ear is the part you see and the hearing canal as far as the eardrum. The inner ear is the snail's shell, the cochlea. And the middle ear is the part between the eardrum and the oval window, where the three tiny bones are. The cochlea is like a kind of head which has a shoulder an arm and a hand touching the eardrum — the three tiny bones. One could say this head, the cochlea listens with its little hand to the vibrations of the eardrum, and that is how we hear.

The cochlea which lies well inside the ear, is surrounded and protected by strong bones. The cochlea is your real ear, where your hearing begins. What comes before — outer ear, hearing canal, the three little bones — only prepares the sound. It was the same with the eye. The real seeing begins in the retina, at the back of the eyeball. What comes before — the cornea, the iris, the pupil, the lens and the jelly — only prepare the little picture appearing on the retina which, like the cochlea, is well inside, surrounded and protected.

It is important that the light and the sound outside are not allowed to come in just as they are, but are prepared before they come to the retina or cochlea. Even the piece of bread we eat is not allowed to get into our blood and muscles just as it is, but has first to be prepared in the stomach and intestines before the villi let it pass through the walls of the intestines. And even the air we breathe is carefully filtered in the nose and throat before it gets into the lungs.

And the nervous system is even more careful than the digestion or the rhythmic system. Only the most refined part of the light of the outside world is allowed beyond the retina where seeing begins, and only the most refined part of the sound from the outside world is allowed to the cochlea where hearing begins.

Our head is like the castle of a king, who will only allow the noblest and most refined ladies and courtiers to be admitted to his presence, a king who keeps himself aloof from the rough and everyday life outside his palace. In real life we may not like such a proud king in his castle, but our head has to be like this, otherwise we could not think or know anything about the world.

Our hand, in which the will works, must touch the real world as it is. But our head needs only a very refined picture of the world. If the full force of the light or the full power of

sound could enter our head then we would be overwhelmed
and be unable to think at all. Sometimes the noise of an explosion is so loud that the mighty vibrations go right through to
the cochlea and people are knocked unconscious. So there is a
great wisdom in letting only refined light go on into the head
from the retina and in letting only refined sound go on beyond
the cochlea.

What happens with the sound in the cochlea? The vibrations of the eardrum are amplified by the little bones and reach
the oval window which is a thin skin on one spot on the outside of the cochlea. The oval window now also vibrates to the
same sound as the eardrum. Inside the cochlea there is watery
liquid. There is a very good reason for having liquid in the
cochlea.

You may have noticed when swimming or at the shore
that sound travels better through water. Water vibrates more
strongly than air. So the vibrations of the oval window now
make the liquid inside the cochlea tremble and vibrate to the
same sound that came from the eardrum.

Amazingly inside the cochlea there are thousands of tiny
hairs of different lengths: some are shorter, others longer. There
are hundreds of different lengths of hair in the cochlea. These
tiny bristles in the cochlea are like little strings, with each length
resonating to a particular sound: the shorter ones to high tones,
the longer ones to low tones. For every sound we hear there are
little hairs vibrating in harmony. The auditory nerve goes from
the cochlea to the brain and carries the message of the tiny hairs
to the brain. There are thousands of tiny strings in this snail's
shell and this orchestra of tiny strings is the real ear, through
which we hear.

From each middle ear there is a tube going to the throat,
called the Eustachian tube. When the air pressure outside
changes — for instance as you climb a mountain or with
changes in the weather — you sometimes feel your ears pop-
ping. The Eustachian tubes let air into the middle ear so the
pressure there is always the same as the pressure outside. The
Eustachian tubes equalize the pressure. When soldiers fire a big
cannon they open their mouths. They do this so the vibrations

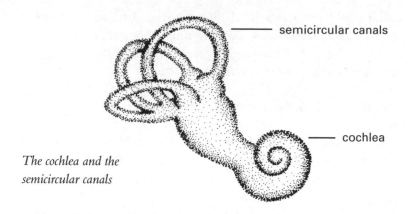

——— semicircular canals

——— cochlea

The cochlea and the
semicircular canals

of the bang reach the eardrum from two sides, the ears and the Eustachian tubes. The terrific air pressure then presses from both sides with equal force, so the sensitive skin of the eardrum is not broken.

There is still another thing in the ear. On top of the cochlea is a rather strange-looking arrangement of hoops, or rather half-hoops. Called semicircular canals, they are of enormous importance for us. These three canals are arranged at right angles to each other. They are hollow and contain a fluid. On the inside there are some little hairs, a bit like the cochlea, but you do not hear with these canals. If you lean forward, back or to the side, the fluid in the semicircular canals moves and the little hairs detect this movement. Through this you know the position of the body and you can walk upright and keep your balance even with your eyes closed.

If anything in these semicircular canals is disturbed, you can neither stand nor walk. If you turn round and round quickly several times and then stand still, the water in these canals continues to go round after you have stopped, and so it seems as if the room is spinning. When people drink too much, the whole sense of balance does not work properly and so drunkards stagger about. These canals are also somehow connected with the stomach: when you are in a boat on rough seas, like the water outside the fluid in the semicircular canals slops about making you feel giddy and sea-sick.

These three semicircular canals help to give us our sense

of balance: without them we could not ride a bicycle, walk or stand. When a little child learns to walk, it learns slowly to follow the guidance of these little canals on top of the cochlea. So our sense of balance is also located in our ear, in these semicircular canals.

Balance, Taste
and Other Senses

Earlier we learnt that we can only see things in three dimensions because we look at the world through two eyes, because we have binocular vision, each eye giving a slightly different picture. What are these three dimensions? In geometry a point has no dimensions — the point you make with pencil or chalk is not really a point, it is only a sign that is supposed to be a point. A point moving forms a line, which has one dimension, its length. When a line moves, it describes a rectangle. Squares, rectangles and circles are areas which have two dimensions: length and width. If a square moves up or down it forms or describes a cube. A cube has three dimensions: length, width and height.

The things we see in the world all have three dimensions, not more and not less. Lines with one dimension or squares with two do not exist by themselves, they are only part of three-dimensional things, and nothing has more than three dimensions.

These are the three dimensions in geometry which we can grasp with our thinking. But our body moves in three dimensions: backwards or forwards — one dimension; left or right — the second dimension; up or down — the third dimension. These are the real three dimensions of our life. As a little child, when we begin to walk, we get to know these three dimensions not in our head, but we *experience* them with our whole body. It is only because we have learned as toddlers to move in these three dimensions with our body, that we can later think and talk about the three dimensions in geometry.

We saw that we can only learn to walk because we have in the inner part of the ear above the cochlea, the three semicircular canals. Just as the eye is the sense organ of seeing and as the ear is the sense organ of hearing, so these three little canals in the ear are the organ for the sense of balance. The wonderful thing is that these three half circles are arranged at right angles to one other. One is for backward-forward movement, one is for right-left movement, and the other one is for up-down movement.

So the three dimensions are already built into our body; they are already there in the inner part of the ear when we are born. When we learn about three-dimensional geometry and volumes we only learn something which is already in us and which we already know from experience.

Let us compare the different senses. With the eyes we can see stars which are millions of miles away. The eyes take us farthest out into the world. The ear, our hearing, does not extend nearly as far: even the loudest bang, the most terrible explosion can only be heard some hundreds of miles away. The sense of smell is closer. Even the strongest smell like the scent of a field of new mown hay, will only go for a mile or two. With the sense of touch something has to be right beside you, within reach of your hands, before you can feel if it is rough or smooth. Similarly our sense for warmth is closer: we only know if the air is warm or cold when it reaches our skin. When it comes to the sense of taste, we have to take something inside our mouth to know how something tastes.

The sense of taste is very interesting. It is not your whole mouth which can taste whether something is sweet or sour, but only the tongue. The roughness on the top of your tongue is made by thousands of little taste buds. They can tell one taste from another. The underside of your tongue has no sense of taste at all.

But it is more complicated: only the taste buds on the tip of your tongue can taste sweetness; if you put a lump of sugar on the back of your tongue you will not sense any taste at all. You taste sweetness only with the tip of your tongue. And on the sides of your tongue you taste what is sour. Right on the back of the tongue are the taste buds for bitter things. And the taste buds for a salty taste are all over the tongue; every part of the tongue can taste salt.

In a way, it is like the little hairs inside the cochlea. In the cochlea one little group of hairs vibrate to one particular tone and not to any other. They are tuned in to one tone only. And in the same way each part of your tongue is tuned in to one particular taste. Human beings can only distinguish four tastes: sweet, sour, salt, bitter. It is possible that animals — cows for instance — can sense more tastes. While we think it must be boring eating grass all the time, for the cow the grass by the riverbank might taste quite differently from the grass up on the hill. But one thing remains the same for the cow, dog, cat and for us human beings: we can only taste something which is liquid and dissolves in the mouth. Anything that does not dissolve at least a little, a stone for instance, has no taste for us. Tasting is the beginning of digestion. When we taste something, it begins to become part of our body, and we can only take into our body what is liquid.

Comparing the senses we began with sight (which takes us far out into the world) and came nearer until with the sense of taste we are inside our body. Then there is the sense of balance — the three hoops — that tells us something about the position and movement of the body. The sense of balance is a body-sense rather than a world-sense. We have still other senses, which are purely body-senses. For instance, we sense if our body as a whole is fit and healthy. At times we may not be really ill, but we have a sense of not being well, and at other times we feel a real glow of health. This is also a sense; a body-sense, not a world-sense. So we have quite a number of senses and not just the five commonly known ones.

We might think that when we look at something our eye automatically does all that is necessary for us to see, or we touch something with our finger and the nerves in the skin automatically tell us whether it is rough or smooth, hot or cold, hard or soft. We might think that all this is a mechanism — a kind of complicated machine — which functions automatically. But it is not so at all.

For instance, Helen Keller was blind and deaf, but by her own will-power she developed such a fine sense of touch that she could enjoy a song by a famous singer by holding her fingertips a little distance from his lips while he sang a song. Her sense of touch was not so sensitive at first, but with practice she developed it. And blind people often develop a much more acute sense of hearing than sighted people possess.

This shows that the human soul is in the senses, and that the soul can even strengthen one or the other of the senses. The senses are not just a dead mechanism; the soul lives in them and can develop and strengthen them. We can also make them weaker. If the soul has no love for beauty in painting or music, the eye and ear become duller, and such a person cannot appreciate the subtleties and full beauty of the world. One could say the soul lives in the senses and educates them as a teacher educates children.

The Brain

At a very young age an illness left Helen Keller blind and deaf. Her whole world was a dark and silent prison. She had lost the two most important senses — sight and hearing — the two main windows through which we look out into the world. And then this wonderful teacher, Anne Sullivan, came. Miss Sullivan could not give eyes or ears to Helen, but she opened another way to know the world, and that was through thinking. Miss Sullivan used Helen's sense of touch to spell words in finger-language into her hands. But the important thing was that the words gave Helen the possibility to think, and once she could think, the world was open to her again.

We know the world through our senses *and* through thinking. We know through our senses that, for instance, there is day and night, that the sun rises and sets. But we must use our thinking to remember that some months ago the days were longer and the nights shorter than they are now. We know the rhythm of the year, the rhythm of the seasons only because we can *remember*, because we can *think*. All our knowledge comes from these two sides: the senses and thinking.

The sense organs and the nerves, which go from the sense organs to the brain, tell you whether it is dark or light outside, but they don't tell you that the sun will rise again tomorrow morning. For this you must think, you must use your brain. Both the brain and the sense organs, as well as the nerves which connect them, belong to the nervous system. Through the nervous system we know the world.

The digestive system gives us heat and energy so that we can move (without it we would be unable to move), but the digestive system gives us no knowledge of the world; we would move about blindly and mindlessly. The nervous system is the opposite: its does not move at all. If there were no heat and energy from the digestive system in the body, the nerves could not move a single muscle. But this nervous system gives us knowledge of the world. The rhythmic system stands between the two opposites and makes them work in harmony with each other.

We have seen that the sense organs are one way through which the soul knows the world. Now we have to learn something about the brain, that part of the nervous system which the soul uses to think and to remember. In fact, thinking begins with remembering. You must first remember the words you have learned to think and then speak your own sentences; you must remember how day and night were back in the summer to think of the rhythm of the year.

Now to understand how our memory works, let us recall the after-image. You look at a red square, look away, and on a blank surface you "see" a green square. But that green square is not real, it is only an image produced by the eye after it has seen the red square. If I draw a cross on the blackboard, you look at it, and then I wipe it off, you can still see in your mind what it looked like. You have a memory-image, a kind of after-image, though in the right colour. But it is also not real, it is only an image.

To have this memory-image you have used your brain. Unlike the after-image produced by the eye, you can call up the memory-image any time you like, even years later. Another difference between the two images is that the eye's after-image appears automatically without effort, while the memory-image made by the brain does not.

It can happen that one of you is not attentive in a lesson, that your mind is elsewhere. Your body is here, your ears hear the sound of my voice, the sound travels through the cochlea, the nerves carry the message to the brain, but because your soul has not been attentive, you cannot remember what was said and then there is no memory image. The brain itself cannot produce a memory image, it is the soul which uses the brain to produce the memory picture.

One could say the soul uses the brain as a kind of notebook. If you make notes of what was said, you don't blame the notebook if something is left out, you can only blame yourself. The notebook will be as good or as bad as you make it. And the brain — although infinitely more wonderful and more complicated than any notebook — is also only as good or as bad as your soul makes it. A clever person has not got a bigger brain than other people, but a more active soul which makes better use of the brain.

The soul can even overcome the difficulties of a damaged brain. For example it can happen that one of the little blood vessels which carry blood to the brain breaks or bursts. In any other part of the body the bursting of a little blood vessel hardly matters, but in the brain it is a serious matter because if any part of the brain is left without blood for a few minutes it dies and nothing can make that part alive again. We call such a bursting of blood vessels in the brain a stroke.

The little area of the brain under the left temple is called the speech centre where the memory of how to speak is located. If by a stroke, by the breaking of a blood vessel, this area is for a short time left without blood and dies, the person can no longer speak. They can understand what you say, but cannot speak. But people who have had such a stroke have often learned to speak again. It takes a long time and patient work over months and years. As they learn to speak again, they form a new speech centre in another part of the brain.

So the soul is master of the brain. With great willpower and patience it is possible to change the brain and built a new speech centre where there was none before. It is not the brain itself which simply grows a new speech centre, but the soul which creates the new speech centre.

From the example of the stroke we can also learn something else. There is no such thing as a stomach stroke; the bursting of a few blood vessels makes very little difference to the stomach. It is also possible to operate on the stomach and cut out as much as half of it, and the remaining half will heal and work well again. Of course, not perfectly, but the patient will be able to eat fairly normally. You cannot do that with the brain. If the tiniest part of the brain is removed by an operation, it is gone for good. The brain does not heal as the stomach does. The power to grow what has been cut is called regeneration. The stomach has great powers of regeneration, but the brain has hardly any at all.

So the brain is the least alive part in your whole body, it is only kept alive by the blood but if it stops for a few minutes the brain dies. The brain is almost without life, but just because of that the soul can use it to know the world. The stomach gives you real things, food, but it cannot give you a picture of the world. The brain gives you nothing real. It is like a dead, faithful mirror, only giving images of the world, but through these images the soul can explore the world, unveil the past and plan the future. This is the most important thing about the brain: for our power of knowledge we use the least alive part of the body, the brain.

The Brain and the Spinal Cord

While the skull which protects the brain consists of the hardest bones of the body, the brain itself is soft. It is not really one brain, but it consists of two halves which are connected by a bridge between them. The brain is round but has folds and looks a bit like the kernel of a walnut. The interesting thing about the folds is that no two human beings have the same folds. When we are born, the folds are only very slight and all brains are similar. But as we grow older using the brain more and more, the folding becomes stronger, more and more folds (and folds within folds) appear. Just as we all make different use of the brain — an artist, scientist or engineer each uses the brain in a different way — so the folds develop an individual pattern.

Each part of the brain has a particular function. We heard of the speech centre. But even in the speech centre, the part for your mother language is not the same as the part which learns a foreign language. Sometimes a stroke victim loses their mother language but can still speak a foreign language he had learned. There are centres for seeing and centres for hearing and centres which know how your arm or your leg moves. If such a centre is damaged, the arm or leg becomes paralysed.

Another interesting fact is that the brain does not simply lie in the skull, it floats. It is surrounded by water, the cerebrospinal fluid. There is a very good reason for this. We saw that the brain needs tiny blood vessels, capillaries to keep it nourished. If the brain were to rest with its whole weight on the capillaries, it would squeeze them, preventing the flow of blood. Just as our own body has less weight in water, so the brain weighs less when floating. It floats in the brain fluid, hardly pressing on the blood vessels which nourish the brain and keep it alive.

Our feet stand firmly on the earth, but our brain floats. Our brain is, one could say, excused from physical work: that is for the muscles and the digestive system. The brain is lifted above such work and floats, hardly feeling its own weight.

The brain is not there to do any physical work; it is there to be a faithful mirror for the mind, for the spirit. No matter how hard you think, you don't move your brain, it does not expand or contract like the heart or the stomach. But something does happen in the brain when you remember or think. Very fine electric currents pass through the brain. One could say very fine lightning is flashing through your brain. It is a very small current or fine lightning which needs delicate instruments making scans to discover these flashes.

Another thing also happens when we think. Tiny crystals form in the brain — one might call it a very fine salt or brain sand — but they are a burden to the brain. So the blood has to dissolve the crystals, and this happens when we are asleep.

So the brain does not move while we think, but there are electrical and chemical processes going on in the brain when we think. Through these electrical and chemical processes the soul works on the brain and uses it to remember and to think.

While the brain is the centre of the nervous system, there is another important part. The spine is not just solid bone. The vertebrae each have a hole which together form a flexible tube. Inside this lies the spinal cord. The spinal cord is a bundle of nerves. At each vertebra fine nerves branch out to all parts of the body. The cord does not go right to the bottom of the spine, but only about as far as your waist. The head can sense what

The brain and spinal cord seen from the side.

is happening in the big toe, because there are nerves which connect the brain through the spinal cord right down the leg into
the big toe.

There are some things which the spinal cord does by itself
without the help of the brain. These are reflexes. If by accident
you touch a hot stove with your hand, the brain will say: "This
is hot, it hurts." But long before the slow brain in the head has
discovered that it is hot and hurts, the spinal cord has said with
lightning speed: "Hot, danger. Hand away!" If you have ever
touched something hot unawares, you will know that you take
your finger away quickly before you even have time to think or
consciously notice it.

You can learn new reflexes. For instance, when you first ride
a bike, to begin with it is your slow brain which has to tell you,
"Now you are falling to the left, do something about it ... now
you are falling to the right, do something about it." And you go
wobbling down the road and sometimes fall off. But after a few
days of trying, your spinal cord learns the new skill and then you
can balance much better without even having to think about it.
Of course you still need your brain to tell you where you want
to go on your bike, or that a dog has run out into the road and
you have to stop suddenly.

It is the same if you learn to play an instrument; you keep
practising your music every day until, after a few weeks, you can
play "automatically." Then your brain can concentrate on play-
ing the music beautifully, without having to think every time,
"Oh, now my third finger has to go into this position." It just
happens without thinking, it has become a reflex.

Just as exercise — walking or swimming — is good for our
health, keeping our muscles flexible and strong, so when we
learn new skills we are exercising our brain and spinal cord.
They too need exercise if they are to stay supple and active as
we get older. And we can go on learning new things all our
lives.

The brain and spinal cord together are known as the central
nervous system. There is another part of the nervous system
called the autonomic nervous system. This autonomic nervous

system also has a centre, a kind of second brain, running down the spine all the way from the base of the brain to the very bottom of the spine.

The nerves which go to the brain from the eyes, ears, nose, tongue and skin tell us about the world outside. But the nerves that go to the autonomic nervous system, in the spinal cord, bring messages about what is going on in the body. Just think how the work of the different organs has to be coordinated. The liver, kidneys, stomach, heart, and lungs all must function at the right moment, otherwise there would be chaos. And the brain which remembers when and what each organ has to do is not the brain in your head but the autonomic nervous system centre in the spine. So we have not only the one brain in the head, we also have a second brain in the spine.

The memories in our brain are our own, but the memories in the spine belong to the whole human race. The memories in our brain are those which we have acquired since we were born, but the memories in the autonomic nervous system were already there at birth. The brain in our head knows nothing when we are born, but goes on learning as long as we live. The spine brain knows all it needs to know at birth, and does not learn anything new in the course of life. The central nervous system is at birth like a book of blank pages, but the autonomic nervous system is a finished book to which nothing can be added.

Of course we need both kinds of wisdom: the wisdom which is my own and which I collect in my brain, and the wisdom which I received as a gift and which uses the spine brain to guide and protect me. But there is a difference between these two kinds of wisdom: I am conscious of the knowledge I have learned myself, but the other wisdom is unconscious.

The brain, spinal cord, nerves and sense organs together form the nervous system. Through the nervous system the soul has images of the world and of its body. These pictures guide, warn and help us; through these pictures we know the world and think.

In the digestive system — muscles, stomach and intestines — we don't have images or knowledge, but handle real things and act, move and change the world: we will.

The rhythmic system is the healer, standing between the two others; through the rhythmic system we feel — we love or hate, we are sad or happy, we hope or fear. We are really threefold beings: threefold in our soul, thinking, feeling and the will, and threefold in our body with the nervous, rhythmic and digestive systems. In the human body the three systems are balanced.

The Threefold Human Being

Our brain, our nervous system, is the least alive part of the human body — it is almost dying every day, but is kept alive by the blood and the digestive system. The blood feeds and nourishes the brain, and of course the nourishment comes from the digestive system. The digestive system is the most alive of the three systems.

In the course of seven years the whole substance of the body, even the bones, is renewed. The bones consist mainly of calcium, and if we do not get enough calcium in our food then our bones become thin and brittle, and break easily. The old calcium is pushed out and has to be replaced by new calcium from our food. Through the digestive system a new body is formed every seven years, or one could say a new body is "born" every seven years.

At the beginning of life is birth, and at the end of life stands death. These moments are the great birth and the great death, but there are also a little birth and a little death. Every day we die a little in the destruction and crystallization in nerves and brain, and every day we are "born" a little in the digestive system. But if we had only the life-giving birth-giving digestive system we could not think or be awake: we would be sleeping all the time. It is the nervous system and brain, the system of the little death which makes us conscious, thinking beings.

Every day the digestive system gives us a little birth, a little building up of the body. And so it is only natural that the great birth, the building of a whole new body, takes place in the digestive system. The third cavity is also the place where before birth the new body of a child is formed.

We enter the earthly life through the digestive system, and we could say we leave the earth through the nervous system. Every time we think, we leave the earth a little, as we will leave it when we die. Every time we use our will, our digestive system, we enter earth-life a little, as we entered it at birth. And our feeling, our rhythmic system, holds the balance between the two.

There are creatures in nature which are not quite at home on earth; they are always rather remote from earth. A bird, hopping on the ground, is not really at home on the ground, it is at home in the heights, where it can soar and fly. The priests of ancient Egypt looked at the eagle soaring in the sky and saw this as a picture of thinking: the wings were a picture of the power of thought. Later when painters painted angels with wings they did not mean that angels needed birds' wings, but that angels have the powers of thought, to be removed from the earth.

The eagle, in fact any bird, has a very poor digestive system compared with other animals. Birds have very short intestines in proportion to their bodies, and digest food quickly, so quickly that they don't get all the nourishment there is in the food. That is why birds have to eat great quantities and never cease looking for food.

The priests of ancient Egypt also spoke of animals which have an excellent digestive system, even better than our human digestive system. These are the ruminants, the animals which chew the cud: cows, sheep, deer. A cow has not just one stomach but four. The cow crops the grass (bites it off) and gulps it down into one stomach, the paunch which is only a store. And when the paunch is quite full it brings the grass up again and chews and munches it thoroughly — that is chewing the cud — and only then does the food go into the other stomachs. Because the food is so well digested, the cow's manure has a nice smell, unlike the droppings of other animals. The cow, which has such a thorough, wonderful digestive system which extracts the best out of every mouthful of grass, is a heavy, earthly creature. It has no inclination to rise above earth, as the eagle has. The cow is the very opposite of the eagle.

The Egyptian priests also spoke of a third kind of animal which has a well-developed rhythmic system. These are the cats: tigers, leopards, lions. The heart, the whole breathing and blood circulation of a lion has more vigour, more strength than that of any other animal. Having such a strong heart, the lion also has the great virtue of the heart: courage. That is why a very courageous king was called Richard the Lionheart.

The priests of Egypt said: eagle, lion and cow or bull are like images of the nervous system, rhythmic system, digestive system, of thinking, feeling and the will. If one could put these creatures together so that they balance each other, there would be the image of the human being. That is why they made figures of a being that had a human head, the wings of an eagle, the trunk of a lion and the feet of a bull. Such a figure made up of man, eagle, lion, bull was called a sphinx.

Muscles and Bones:
Anatomy

Calcium and the Bones

The sea is the great reservoir of the water on which all living things depend. Scientists believe that life not only depends on the sea, but began in the sea. There is a reason for this belief. A chicken, for instance, is quite a complicated creature with heart, lungs, muscles, bones, feathers. But this chicken is at first simply an egg. And it is the same with all the higher animals and with the human being: they begin life as an egg in their mother's body. But in the sea there are creatures, amoebas, which float in the sea all their lives and are little lumps of a substance like egg-white. It is thought that in our own development we repeat in a short time all the stages which life took over millions of years, but that all life began in the ocean.

Two thirds of our body are liquid, salt water like the sea, as if this were another reminder that life has its origin in the sea. We have left the ocean and live on dry land, but we carry a little ocean in our blood, in our body, with us. There is a great difference between the liquid and the solid parts of the body. In physiology we heard about the wonderful work done by the blood. It carries oxygen to all parts of the body and carbon dioxide back to the lungs; it carries the food from the intestines to all parts of the body, it heals wounds and sores and it produces antibodies against germs. We can easily see that it is the blood, the liquid in us, which makes us living beings. The life of our body is in the liquid part, in the blood.

It is the same with plants, with trees for instance. The life of a tree is in the sap, in the liquid which rises from the roots up to every branch and leaf. In winter the sap stops rising and the tree is temporarily as good as dead. The hard wood and the bark have no life of their own, they are dead without the life-giving sap. Similarly the hard, solid parts of the human body, the bones, have little life of their own. They are kept alive by the blood, by the liquid part of the body. This is the real and important difference between the liquid and the solid parts of the body, between blood and bones. The liquid part of the body has life, but the bones are comparatively dead.

And because the bones are dead compared with the blood, people have always regarded the bones, the skeleton, as a picture of death. On bottles containing a poisonous substance there is often the skull and crossbones to warn people that the contents of the bottle can bring death. Pirates hoisted the Jolly Roger, a flag with the skull and crossbones, to show that their intention was to kill those who fell into their hands. In the Middle Ages painters showed a skeleton leading kings, bishops and knights away from throne, palace and castle. This was to remind people not to become too attached to earthly goods, because death would take them away. They should desire treasures of the spirit — kindness, honesty, wisdom — which remain with the soul even after death. These painters too regarded the bones, the skeleton, as a picture of death.

At that time there were also the first modern scientists, who made experiments in physics and chemistry. Called alchemists, they knew something which we also heard in physiology. The blood in our bodies does not remain the same; it is continually renewed. The place where the new blood is produced is inside the bones. The bones are hollow, and contain marrow. It is in that marrow that the living, new blood is produced.

It is one of the wonders of the body. Inside the hard skeleton — the picture of death — new life is born: our living new blood is continually produced there. These early scientists, the alchemists, knew this and encapsulated this knowledge in a Latin verse which, in translation, reads:

Behold the man of bone — and you see death
But look inside the bones — then you see the awakener of life.

The real meaning of the verse was "as it is in the body, so it is everywhere — death is not the end, it is the beginning of a new life. Learn from the wonder of the body that in death a new life begins."

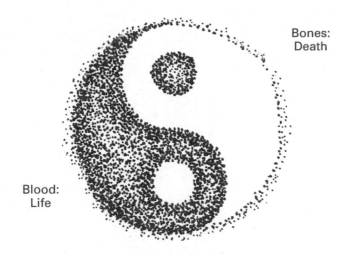

Bones:
Death

Blood:
Life

Blood and bones — these opposites — renew each other. One could put it in a picture as shown above.

We can see how these opposites work together. The living blood feeds and renews the bones all the time. Our food contains calcium and our bones are renewed all the time by calcium. If there was not enough calcium in the food, the bones would soon become thin and brittle, for the old calcium is slowly taken away by the blood and must be replaced by new calcium. Much of the calcium we need for our bones comes from dairy produce, like milk, cheese, yoghurt as well as from green vegetables and bread. Babies and young children need milk to build their growing bones. It is the blood which brings the calcium to the bones and so constantly renews them. But the bones renew the blood: new blood is produced inside the bones.

Calcium is the great builder of the world. Corals which build great reefs and atolls use calcium from the sea. The oyster and all shellfish build their shell from calcium; the mother-of-pearl you see inside some shells is calcium and so are the pearls themselves. Eggshells are also calcium. Limestone mountains have been built of calcium from such sea creatures over millions of years. And human builders use calcium (lime) to make quick lime which they mix with sand to make cement and concrete. We cannot build without calcium.

It is also calcium that is used to build our bones. The solid part of the body consists of the same material as we find in limestone mountains, in corals and seashells, in a precious pearl necklace, in the gleaming white marble of a Greek temple or statue.

The Structure of Bones

Looking at the contrast between flowing, living blood and the solid, comparatively lifeless bones, we must not forget that without bones we would be soft shapeless creatures that could only exist in the sea. To live on the solid earth, we must take into our body some of the hardness, some of the solid strength of the earth to make our bones.

But at the beginning of life we are still soft beings. The embryo in the mother's womb is at first a very soft little creature. Slowly certain parts in this soft little body begin to harden a little, becoming cartilage. This not quite hard cartilage is an important substance, for some parts of the body remain cartilage. The ear is cartilage, and there is also some cartilage connected with our bones once they have completely hardened.

In the embryo all the bones are at first the soft, elastic cartilage. When the baby is born, the bones are a bit harder, but not yet really hard. You only have to touch the skull of a very young baby to feel how much softer it is than your own. It takes about one and a half years before the bones have become hard and solid, and that is the time when the toddler begins to stand up and walk. Before that time the bones would not be strong enough to carry the weight of the body. So babies need a lot of milk for the calcium to build their bones.

The bones not only grow harder, they also grow in size. And this is by no means simple. The bone grows in length as well as in thickness, but it grows more in length. When you are fifteen years old, the bone in your upper arm (the humerus) is about three times the length it was at birth, but in thickness it has not grown all that much — it is not three times thicker than a

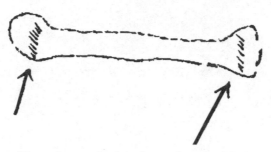

The main growth takes place at the ends of the humerus

baby's. The bone grows because the blood brings new calcium to it, but how does the body know where to put the new calcium? And another question is this: How does the body know that it must give most calcium to the limb bones and less to the skull bones? While in the first fifteen years the humerus grows to three times its length, the skull grows only a little.

The humerus, the bone of your upper arm, begins to grow in a remarkable way. Near each end of the bone there is a part which remains cartilage much longer than the rest. From these ends, the main growth in length takes place.

But during growth the whole bone is also shaped and remodelled. While we are growing — and the skeleton is not fully-grown until we are about 25 — the bones are not just getting bigger, they are partly dismantled and reconstructed. Yet all the time we have a complete bone we can use. We human beings cannot change part of a machine while the machine is in use, but the amazing fact is that the unconscious wisdom in the blood can do it.

The wisdom which builds our bones is also an engineer who knew certain laws of mechanics long before human engineers discovered them. The ancient Greeks and Romans used thick solid pillars to carry the weight of their temples. Instead of such heavy pillars modern engineers use hollow tubes, which are thinner and lighter than solid pillars. Our limb bones would be terribly heavy if they were solid, but they are not; they are hollow tubes. The wisdom that made the bones knew the hollow tubes principle millions of years before we discovered it.

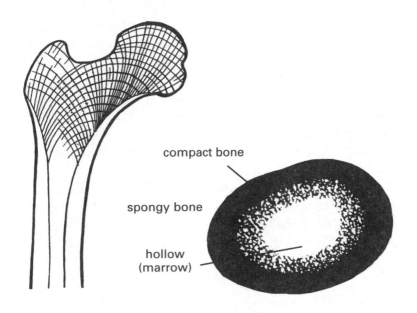

compact bone

spongy bone

hollow
(marrow)

Left: Spongy bone structure inside the head of a human femur (after Wolff)
Right: Cross section of a bone showing the three parts

And now we come to the tube itself — the substance of the bone around the hollow. If you look at an old metal bridge, you can see a criss-crossing of metal pieces. These ties and struts are the means of distributing the weight and making the structure rigid. A solid bridge or a solid crane would not only be much heavier, but could carry less weight than one made up of ties and struts. But it needs quite a bit of mathematical knowledge to work out the arrangement of girders for a bridge or crane.

On the inside of the walls of the bones there are thousands and thousands of tiny ties and struts arranged so perfectly that no engineer could improve on it. That is why the light bones we have can carry great weight without breaking. Every tiny strut on the inside of our limb bones has just the right shape and size and position to give the greatest strength with the least amount of material. Through this arrangement, the bone can resist pressure from all sides.

If we cut through a bone, the cross-section shows at least a glimpse of the intricate pattern of the bone. On the outside is the hardest, most solid part of the bone, called compact bone. Then on its inside comes the so-called spongy bone which is the part with the tiny ties and struts and the little empty spaces between them. And inside this spongy bone is the hollow.

So the bone is really threefold: compact, spongy and hollow. The compact and the hollow are opposites, and the spongy bone stands in the middle. This threefoldness — two opposites and the third between the opposites — is something we find again and again in human body.

So the growth of the bone is not at all a simple matter, and even what we have just seen is only a small part of the wonders of every bone of our body.

The Joints of the Bones

The growth of a single bone such as the humerus of the upper arm is a wonderful process, and the structure of such a bone is a masterpiece of mechanical construction. We saw the three parts: outside the compact bone, then the spongy bone, and in the middle, the hollow part of the bone. The compact bone and the hollow are opposites, while the spongy bone is neither quite solid nor quite hollow, it is between the opposites. This pattern is found again and again in man and in nature: two opposites and between them a kind of balance as a third.

The bones of the whole human skeleton can be divided into three groups; two are opposites and the third group stands as a kind of balance between the opposites.

Let us begin with the bones of the human head, the skull. The round dome of the human skull is not just one round bone, but is made up of eight different bones. There are another fourteen bones of the face.

In a very young baby these separate bones of the skull are not yet quite joined together and there is an open space between them at the top of the head, called the fontanelle. If you gently put a finger on the fontanelle of a little baby, you can feel the blood pulsing through the skin. But as the baby grows older, the bones of the skull grow closer until they are completely joined together. The joints of the skull bones are quite different from any other joints in the body.

If the edges of the skull bones were simple and straight, then with a hard fall or a blow the bones could shift or slip possibly exposing or damaging the brain. To avoid this, the edges of the skull bones are serrated like the teeth of a saw. Like the pieces of

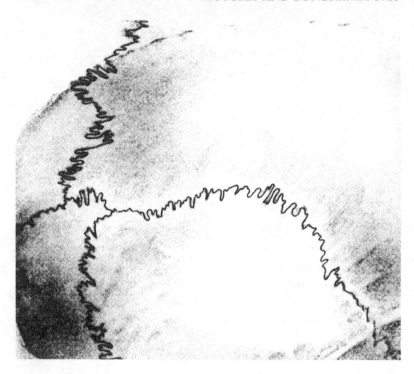

The serrated joints of the skull

a jigsaw puzzle, the bones of the skull interlock and fit together exactly. When they have grown together these serrated edges grip each other so perfectly that the eight skull bones are like one dome-like bone.

The skull bones are not meant to be moved, and neither is the brain which they enclose and protect. The brain does its task by resting quietly without motion inside the helmet of the skull.

But if the bones of our fingers or arms were joined together like those of the skull, we would be completely stiff. The joints of limb bones are quite different. Where we bend our fingers, elbows or knees we find these limb joints are built for movement.

If the hard compact parts of two bones would rub against each other, the bones would grind each other. So at the joints of two limb bones there is a covering of tough, elastic cartilage. This layer of cartilage is shiny and smooth, preventing any grinding of bone against bone when we move.

When the hinges of a door get stiff or rusty, we lubricate them with a little oil to reduce friction. Similarly the joints of our limbs have a watery lubricant. To keep this lubricating liquid in place and prevent it from flowing away, the joint is enclosed by a special tough skin or membrane that holds the liquid.

Just looking at the joints we can see that the skull bones and the limb bones are opposites. The joints of the skull bones are made to prevent movement, while the joints of the limb bones are made to allow smooth and easy movement. Of course skull bones and limb bones are also opposites in their shape. The skull bones are rounded to form a sphere. The limb bones are straight.

The skull bones and the limb bones are opposites in their shape as well as in their joints. The skull bones are built for rest, and the limb bones for movement. The skull bones protect from outside, the limb bones support from inside. And in the middle, between these two opposites, there is a third group of bones, the ribs.

The ribs are curved bones going from the breastbone, the sternum, at the front to the spine. In front the ribs have grown onto the breastbone like the skull bones grow together but not quite as firmly, for the ribs are connected to the breastbone by cartilage. But at the back the ribs are connected to the backbone by joints like those of the limb bones. So as regards joints, the ribs in front are more like skull bones and at the back are more like limb bones.

If you hold your fingers to your lower ribs at the back and breathe deeply, you can feel your ribs move. They are flexible and can be moved a little. So they stand in the middle between the opposites.

This fact that our ribs are flexible is of the utmost importance, for otherwise we could hardly breathe. Your whole chest is like bellows. The lungs themselves have no power or strength to suck in air. A large sheet of muscle, the diaphragm, which lies under the ribs, pulls down and at the same time the ribs lift upwards and outwards (using the intercostal muscles). This expands the lungs to the front and sides, like the expanding of the bellows, and air rushes into the lungs. Then the diaphragm

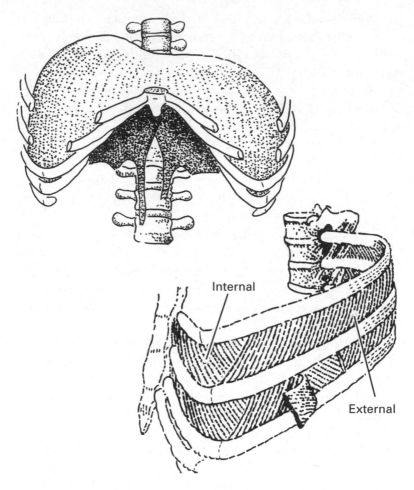

Top: The diaphragm   Below: The intercostal muscles

and ribs relax, moving back, and air is pushed out of the lungs. The movement of the diaphragm and ribs makes breathing possible.

As far as this breathing movement goes, the ribs are like limb bones. But together the ribs form the ribcage which protects the lungs and the heart, just as the skull bones protect the brain. The ribs are a little of both the opposites.

In physiology we heard of the threefold human being and of the three systems — the nervous, the rhythmic and the digestive or limb system. We can see that the bones are similarly three-fold. The skull bones protect the brain, the centre of the nervous system. The rib bones help the lungs, the rhythmic system. The limb bones of course support the limb system.

System	Head	Chest	Limbs
Function	Protects from outside	Protects (heart) and supports (intercostals)	Support from inside
Joints	Prevent movement	Allow small movement	Give free movement
Form	Round	Curved	Straight

The Spine and Vertebrae

There is one group of bones we have not discussed: the bones of the spine. What is their place in the threefold pattern? The bones of the spine contain something of each group — the opposites and the middle — as we shall see. One could say our head, the ribs and the limbs are all suspended from the spine. The spine is really the centrepiece of the skeleton. It is also the spine which makes it possible that we stand straight and upright.

If our spine were just one long straight bone we would be able to stand straight, but we would be terribly stiff. (Just try to keep your back stiff and move your head and limbs at the same time.) On the other hand if our spine were made of rubber, we could bend and wiggle wonderfully, but such a spine could never hold us upright.

So the spine must be able to do two opposite things at the same time. It must support the weight of our head and trunk like a strong rod, and it must be able to bend and move like rubber. The wonderful thing is that the spine *can* do both things because it is neither one hard bone nor a piece of rubber, but is composed of many bones. Altogether there are thirty-three, and twenty-four of these are jointed in the most marvellous way.

These bones which together form the spine are called vertebrae (singular: vertebra). Each vertebra is beautifully jointed with its neighbours. No engineer and no artist could devise a better way to make our spine both strong to carry weight and supple for movement.

Between each vertebra there is a round flat piece of cartilage, a disc of cartilage. If there were no cartilage between the vertebrae then every step and every movement would jar the brain.

These discs are shock absorbers. So the spine not only supports the head, but protects the brain from the shocks and jolts of our movements. If a disc becomes dislocated, you suffer from a slipped disc which presses on the nerves of the spinal cord.

When we look at the spine supporting head and trunk, but moving and bending, and when we look at the bones of the spine as separate and connected by joints, what kind of bones are we talking about? They are the limb bones, of course, for the limb bones are separate with joints, and they support. But this is only half the story.

In physiology we saw that as well as the brain there is the spinal cord which controls all our unconscious functions like when the atria and ventricles of our heart should open or close, or when food should go from the stomach to the intestines. This second brain, the spinal cord, is embedded in the spine, as the brain is embedded in the skull. And as 12 pairs of nerves go from the brain to the eyes or to the limbs, so 31 pairs of nerves go from the spinal cord to the organs of the body. The bones of the spine surround and protect the spinal cord just as the skull surrounds and protects the brain.

So the bones of the spine, the vertebrae, form a kind of skull, but at the same time, they are also limb bones which support, move and have joints. A vertebra is a skull bone, a limb bone and — as we shall see — a rib bone all at the same time. How is this possible?

Let us look at a single vertebra. The vertebrae are all built to the same pattern which clearly shows three different parts.

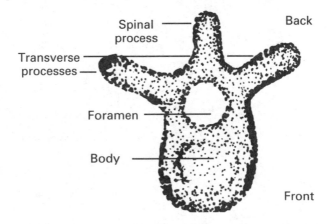

A vertebra

There is one short, thick, solid, straight part called the body. This body is quite clearly a short kind of limb bone made to support weight. The cartilage discs lie between these bodies of the vertebrae.

Then there is a ring of bone. The spinal cord, our second brain, goes through the hole (foramen) of this ring. This round bone, this ring of bone, is the protective skull of the vertebra. These rings fit onto one another and so form a continuous tube for the spinal cord.

Thirdly, there are three small protruding bones called processes. Two transverse processes are at the sides, and the spinal process at the back. These processes are akin to the rib bones. The transverse processes support the ribs where they are jointed to the spine. The spinal processes make up the knobbly part you feel when you touch the spine. So each vertebra itself is threefold: the body is a limb bone, the ring is a kind of skull bone, and the processes are rudimentary rib bones.

This is the general pattern of a vertebra. The pattern changes according to location. In the upper part of the spine the most prominent feature is the ring, the skull part of the vertebra.

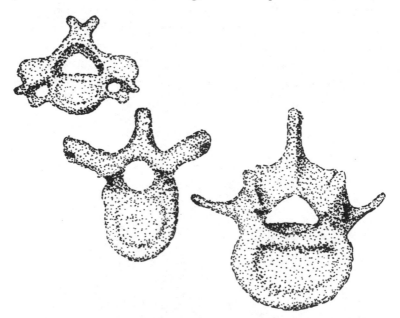

Left to right: Cervical, thoracic (or dorsal), and lumbar vertebrae

In fact the topmost vertebra is only a ring without any body and hardly any processes. The scientific names for these are the cervical (neck) vertebrae. In the lower part the body, the limb part, dominates. These lower vertebrae have enormous bodies and only a relatively small hole for the spinal cord. These are the lumbar vertebrae. In the middle part of the spine, body and ring are balanced, and the processes are especially well developed as the ribs are attached to the transverse processes. These are the dorsal (or thoracic) vertebrae.

So the spine itself is threefold. The upper (cervical) vertebrae are akin to the skull bones and the ring is prominent. The lower (lumbar) vertebrae are akin to the limb bones and the body is prominent. In the middle (dorsal) vertebrae which are akin to the rib bones, the processes are prominent.

The Spine and Balance

The spine has three groups of vertebrae: the hollow cervical at the top, the strong lumbar, and the dorsal or thoracic vertebrae in the middle. The top vertebra, the smallest, is the atlas which

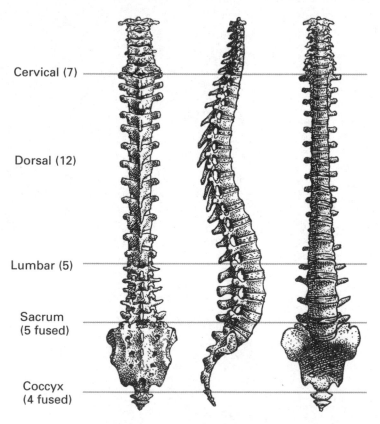

Cervical (7)

Dorsal (12)

Lumbar (5)

Sacrum (5 fused)

Coccyx (4 fused)

The spine from back, side and front

is so jointed to the head that we can nod. Below it is the axis with a little pivot which makes it possible to turn our head to the side. There are seven cervical vertebrae, and below those are twelve dorsal (thoracic) vertebrae, which are jointed with the ribs. Below these are the five lumbar vertebrae which have large bodies and only small rings. The processes of the lumbar vertebrae are fairly large because strong back muscles are attached to them.

The lumbar vertebrae are very important for human beings, far more important than they are for animals. Animals standing on four feet have their weight supported in four places, and are in stable equilibrium. However, human beings have unstable equilibrium; we have to balance our body so that the centre of gravity lies above our two feet, the point of support.

If a man carries a heavy bag on his shoulders, the centre of gravity must lie vertically above his feet, so he must lean forward. If he carried the same heavy weight in his arms, he needs to bend so far back that he could hardly walk. Some people carry even the heaviest loads on their heads; that is the simplest way to ensure that the centre of gravity lies vertically above the feet. But even without carrying any load, the body itself has a centre of gravity which, if we stand straight properly, lies vertically above the lumbar vertebrae. If we stand in a bad posture, then the centre of gravity

is not above the vertebrae, but above the stomach and intestines. However, doing this strains the back and abdomen muscles, our breathing is hampered and the soft digestive organs suffer.

The lumbar vertebrae are most important, for it is with their help that we can stand and walk and keep the unstable equilibrium of the body.

Another thing essential to our
equilibrium is the fact that the
weight of one side of the body is
the same as the other. The bones
of the skeleton are symmetrically
arranged, though the soft parts of
the body are not quite even: the
heart is not in the middle, the liver
is on one side and the stomach is
not quite in the middle.

All creatures with a spine have this symmetry — fish,
amphibians, reptiles, birds, mammals — as well as insects and
most crabs. Where there is symmetry, there is an axis of symmetry, the line along which things are symmetrically arranged.
And in the body the spine is the axis of symmetry. Each vertebra
is itself beautifully and perfectly symmetrical, and the skeleton
is so arranged that each bone on the right has its counterpart
on the left of the spine. This symmetry is not only beautiful,
it is also necessary for our balance to have the centre of gravity
within the spine, not to one side of it.

As well as the one axis of symmetry, we human beings have
another imperfect symmetry. If you compare arms and legs, our
fingers are not like our toes, yet they are related. The hand is not
like the foot, yet they are related, as are arms and legs. One is the
counterpart of the other. It is as if there was originally symmetry, the same pattern above and below, but something changed
the symmetries. Just as the upper vertebrae are different from
the lower vertebrae but show the same pattern, so the arms and
hands are made more for skill, and the legs and feet for carrying
weight, but they come from the same pattern. It is a symmetry
that has been transformed. There is a common pattern; we can
speak of a correspondence.

Similarly above there are the shoulder bones and shoulder
blades, which are very different from the strong and heavy pelvic bones below, but they are again related, coming from the
same pattern. One is the counterpart of the other.

The human skeleton (after Rohen)

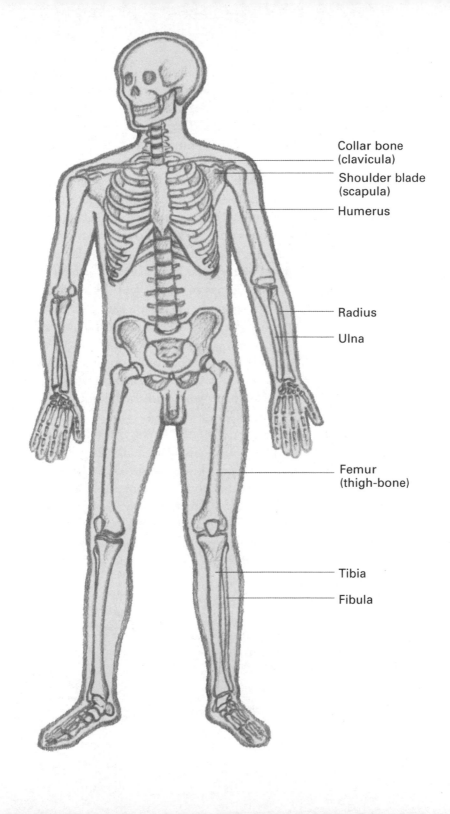

Collar bone
(clavicula)

Shoulder blade
(scapula)

Humerus

Radius

Ulna

Femur
(thigh-bone)

Tibia

Fibula

At the top of the spine there is the head, where the skull bones have grown together to form a beautiful dome. This too has a counterpart. At the bottom of the spine there are vertebrae which have grown together, fused to become one bone called the sacrum. This sacrum is the counterpart of the fused skull bones.

Below the sacrum there is a funny little bone called coccyx (pronounced *koksiks*) which seems to have no rhyme or reason for its existence. For human beings as we are today, it has no purpose, but it is of some importance for dogs, for instance. For a dog the coccyx is the beginning of the tail, and the dog uses its tail to show its feelings. We human beings show our feelings in our faces. We carry our feelings upwards, and the coccyx, that funny little bone, is like a reminder of what we have left behind: a tail to express feelings. So there are counterparts above and below, even with the coccyx.

We shall look at this in more detail. In the upper arm there is a single strong bone, and similarly in the thigh there is a single strong bone. The bone in the upper arm is called humerus, and the bone in the leg is called femur or thigh-bone.

Further down there are two bones. In the arm they are called radius and ulna, in the leg fibula and tibia. In the arm the radius lies above the thumb, and the ulna above the little finger. The name radius tells you that this bone has something to do with going round; it is like the part of a pair of compasses which holds the pencil. The radius can be turned around the ulna, and that is what happens if we turn the palms of our hands up or down.

With our palm turned down, we give, and with the palm turned upwards, we receive. And that we can do this is due to the fact that we have two bones in the lower arm and that one of them, the radius, can rotate around the other, the ulna. In this way two rigid, immobile bones produce a turning motion. The ulna is the stronger of the two and it has

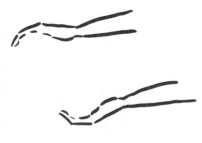

a hinge joint with the large humerus above. This joint is the elbow, and the hard bone you feel there is a part of the ulna.

The counterpart of the radius, the fibula, can hardly rotate around the tibia or shin bone, and there is no need for it. Our legs are much more specialized for the one task of carrying the body. As we neither give nor receive with our feet, there is no need to turn the sole upwards. So the two bones are strong, and the fibula does not rotate.

Another difference is in the joint with the large single bone above. Unlike at the elbow, at the knee joint there is a knee cap, a special bone which makes the bending of the knee a smoother movement because it improves the leverage of the leg muscles. The knee cap makes walking a smooth movement.

Another difference between elbow and knee is that the joints work in opposite directions. The arms bend forwards at the elbow, the legs bend backwards at the knee. And this is just as it should be, otherwise it would put the weight of the body in the wrong place, behind the feet, which carry the weight. Storks and ostriches have to walk this way, and in their case nature put most of their weight well in front of the joint (which is actually not a knee but a heel) so that the centre of gravity is vertically above the feet. Our bodies are not built this way and we could not balance ourselves on such a joint.

We now come to wrist and ankle. Here there is also similarity, but again the ankle bones are thicker and stronger, specialized for their task. The hand and foot, the fingers and toes, clearly show the same thing. Hands and fingers are capable of a thousand and one tasks; they are not specialized for a particular job. But feet and toes are wonderfully specialized for one task: to take the whole weight of the body and to be able to walk.

So the difference between the two parts is that the upper is unspecialized, while the lower is very much so. The upper is free, the lower is unfree and can only do one thing. One could say the lower part sacrifices itself, taking the hard task, so that the upper can be creative.

More Correspondences

The left-right symmetry of the human body is very important for our balance. The correspondence between above and below shows the different tasks of our upper body and our lower body.

The upper arm bone, the humerus, shows the original design better than the corresponding femur of the thigh. It is of course a hard solid bone, but in its shape there is something that reminds us of the flow of water. Water in a riverbed does not just flow on, it rotates and eddies, so the water which is on top now, will be at the bottom a few hundred yards further down. If we could follow a single drop on its path, we would find that this path is a spiral. And now looking at the humerus we see that in the way it is built, there is a twist, a slight turn, the hint of a spiral. There is another similarity to flowing water; where the flow is dammed, the water will spread over the banks on either side, and the river widens. This can be seen at both ends of this bone where there is a thickening and widening as if the bone material were dammed up. The bone shows that it is built by invisible forces of which we know as yet very little.

The humerus

There is another correspondence between lower and upper bones which is not quite so obvious: the hip and the shoulders. The hip bones and the shoulder bones each form a kind of girdle on which the limbs are attached to the body. The arms are attached to the shoulders and the legs are attached to the hips. In anatomy the shoulders are called the pectoral girdle, and the hips are called the pelvic girdle.

The difference between these two girdles is much greater than the difference between arms and legs. The bones of the hips are built for strength and are heavy and massive. The bones of the shoulders are light and free, even slightly transparent. The shoulder blades almost look like wings, while the hip bones are more like a stout seat or throne.

We have come across such a contrast in the small vertebrae above and the solid, heavy vertebrae below, and also in the light bones of the arms and the strong bones of the legs. The same contrast is found between shoulder blades and hip bones, but the difference goes further: the bones forming the upper girdle float apart, and are separate, one on each side, but below in the hips, the heavy bones grow together and form one solid bone, the hip bone or pelvis.

The pelvis has grown together with the lower part of the spine, with the sacrum, and so at the lower end of the spine there is really a solid ring of bone, a throne on which the upper body rests. By contrast each shoulder blade or scapula is quite separate from the spine as the shoulder blades take no part in carrying the weight of the body. Of course the shoulders can carry weight, but the hips *must* carry weight. Projecting from the back of the shoulder blade is a small wing-shaped part, the acromion. This strange bone coming from the back meets another, separate bone, coming from the front of the body, the collarbone or clavicle. The collarbone is jointed to the breastbone in front. The collarbone and scapula together form a kind of loop, which makes it possible for the muscles attached to the bone to lift the arms sideways. If there were no bones to hold and support them, the muscles could not work. Some animals — horses and cows for instance — have no collarbones and so cannot lift their legs sideways.

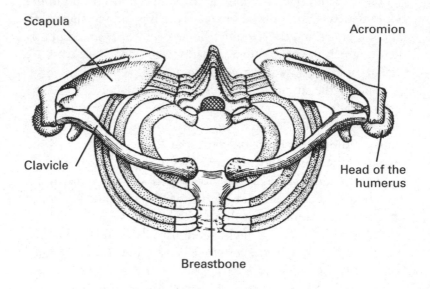

Scapula

Acromion

Clavicle

Head of the
humerus

Breastbone

The shoulder (pectoral) girdle from above

The two girdles — shoulders and hips — are very different, but in one thing they show the common design, and that is the round hollow, the cavity where the limb bones (humerus and femur) are jointed to them. This kind of joint is quite special. We have heard of a pivot joint (which we use when we turn our head), and of a hinge joint (like elbow, knee). This special joint which makes it possible to turn in any direction is a ball and socket joint. It is where the humerus fits into the shoulder and where the femur fits into the pelvis.

Carrying the Weight
of our Body

When we look at the bones of a skeleton, we see something which, quite apart from everything else, is a supreme work of engineering. Let us take a simple example of the work of engineers: an arched stone bridge. An arched bridge is stronger than a flat one. On the arch, the weight pressing down on one point is distributed over the whole bridge. It is the principle of sharing the burden.

The centre-stone is the most important one of the bridge, because it is here in the middle where the weight is farthest away from the ends. The keystone distributes the weight equally to both sides, spreading over the whole bridge.

Similarly the pelvis carries the whole weight of the body. The pelvis has fused with the lower part of the spine, the sacrum. The whole weight of the trunk, arms and head bears down on the sacrum which acts like the centre stone of the

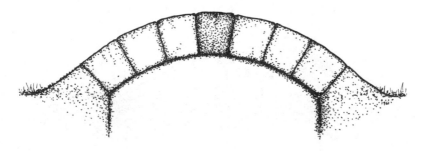

An arched bridge with the keystone at the centre

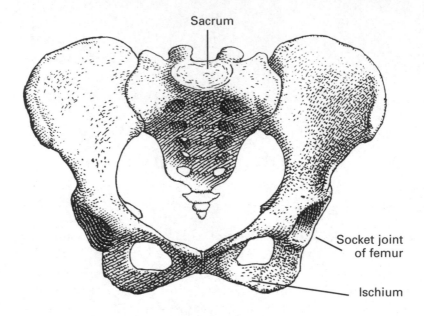

Adult female pelvis

bridge distributing the weight equally over both sides of the pelvis, and then each thigh-bone takes half of the total weight. So the pelvis is a kind of bridge resting on two pillars, the legs. Bridges built by human beings are ones we can walk on. But the hipbone or pelvis is a "bridge that walks," as a book on the mechanics of the body described it. When you sit, the lower arch of the pelvis (ischium) carries the weight. So when we sit the weight of our body is also carried by an arch.

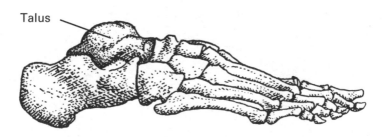

Bones of the foot

But standing, the weight of the body is passed on by the pelvis to the femur (thigh-bone), then through the tibia and fibula (shin-bone and calf-bone) down into the foot. Here again is an arch with a centre-stone, the talus, which distributes the weight over the whole foot. In fact the foot is composed of 26 bones and the weight is distributed in a series of arches. These bones not only carry the whole weight of the body, but can move it swiftly.

Once we come to movement, the structure of the skeleton appears even more wonderful. The femur (thigh-bone) for instance is jointed by a ball and socket joint to the pelvis. The upper end of the femur is almost spherical and fits perfectly into the cuplike cavity of the pelvis. Powerful and elastic cords called ligaments hold the thigh-bone in place. But these cords are very loose for if they were tight we could not move our legs freely. You only have to stand on a chair and let one foot swing backwards and forwards to realize how freely and easily the leg moves.

But if the round head of the femur is not held tightly in its place by the ligaments, what keeps it in its socket? It is held in the cup by suction or air pressure.

The joints are enclosed in a skin or membrane containing a lubricating liquid but no air. The air pressure from outside is so strong that it keeps the femur firmly in its suction cup. (The air pressure is so strong that where the whole leg has to be amputated it takes a very hard pull after all the ligaments have been cut to move the thigh-bone out of its socket.) Scientists only discovered atmospheric pressure about three hundred years ago, but the wisdom that built the human body used air pressure to keep the thigh-bone in its joint hundreds of thousands of years ago.

Walking

Walking is so complicated that if we had to learn it like we learn algebra, we would still be crawling on all fours. But fortunately, just as we eat without knowing the processes which turn the food into strength and energy, so before we learn to think, we learn to walk instinctively, guided by a higher wisdom than our own. Walking is so complicated that no engineer — and a good many have tried — has been able to construct a mechanical human figure which could walk as we can on two legs. Every time we take a single step, about 300 muscles in our body come into action, and the various groups of muscles must act at exactly the right moment. It is a matter of split-second timing, and it all takes place without our even thinking about it.

A step begins with lifting the heel of one foot. At this moment when the toes are still on the ground but the heel is already up, the whole weight of the body is transferred to the other (stationary) foot. So our body is actually unbalanced, one could say we are just about to fall towards the side of the standing foot.

At the next stage the imbalance gets even worse: for now the knee of the moving foot bends, and the whole foot is off the ground, and then swings forward. And then there is an especially difficult moment: the moving foot swings a little backwards and, in doing so, the heel comes to the ground. In this instant — when only the heel of one foot is down — the heel of the other foot comes up so that for a moment we stand only on the heel of one foot and on the toes of the other. Just at this moment, the whole weight of body tips forward and to the other side. We are, as we walk, continuously losing our balance and recovering it. As one scientist said, "Walking is a perpetual falling and recovery." The body really falls from one leg onto the other.

Walking is in fact a dangerous as well as a violent motion. You find out how dangerous it is if you slip and totally lose your balance. And you find out how violent it is if you walk into a lamp post or into another person.

The shifting of the weight of the body from one leg to another — losing balance and recovering it — must be done just in the right measure; too much and you would fall forwards, too little and you fall backwards. No wonder that we start practising early: it is almost the first thing we learn in life.

Just as walking means falling and recovering, so we go through life making mistakes or even doing wrong things, but we recover and learn from the mistakes and wrongs. The whole of life is like walking, almost the first thing we learn.

But to return to walking and the moment when one leg swings freely and the other leg stands. The standing leg is not idle, because at this instant the whole weight of the body has to be balanced on this one leg. But the weight of the body rests on the round head of the thigh-bone which fits into the socket of the pelvis. While the weight is balanced on the thigh-bone, the round knob in the cavity would wobble if there were not fifteen muscles pulling from different directions like the guy-ropes on a tent pole. These fifteen muscles keep the femur, the thigh-bone, in place.

At the moment we balance on one foot, the centre of gravity of the body has to be exactly above the thigh-bone. If the body sways the least bit off this plumb-line, the right set of muscles round the hip joint pull just a bit more and bring the body into the right line again. As soon as the swing of the free leg has come to an end, the body must be tipped forward — it can no longer remain poised over the standing leg — and the muscles of the hip joint now have to meet this change too — and work against the balance.

The whole pelvis is also involved while we are walking. The weight of the upper body rests on the pelvis which is first shifted to one side, over the standing leg, and then not only to the other side but also forward. So in walking we must swing the pelvis forward and from side to side.

Above the pelvis is the spine consisting of vertebrae. As we walk, practically toppling from one side to the other, there are muscles which keep the vertebrae balanced all the time.

But this is not yet the end of the story. As we walk, we swing our arms. This swinging of the arms is always in the opposite direction to the swing of the legs. When our left leg swings forwards, our left arm swings backwards. This is again to help our balance.

We can now understand why about three hundred muscles come into play when we walk: muscles in our feet, legs, hips, along the spine, the arms, right up to the head. It is not only the legs which walk, the whole body takes an active part in it. That is why a good walk is a healthy exercise for the whole body.

Let us look at the work going on in the legs while we walk. When someone turns a cartwheel, the four limbs of the body, the arms and legs, form the spokes of a wheel and the hands and feet are parts of the rim. It is really a rolling round like a wheel with four spokes. Now in ordinary walking we do exactly the same, except that our wheel instead of having four spokes only has two — the legs. Each spoke, each leg, is taken up as soon as it is used and carried forward, but the movement of the legs as a whole is the rolling movement of a wheel.

And the feet are parts of the rim of the big wheel we make when we walk. Now the feet are not curved like a wheel, they are rather opposite in shape, but the elastic muscles make the foot a rolling rim. We come down first with the heel. As the toes come down the heel goes up again and we leave the ground from our toes. From the first state to the last the foot has made a rolling movement: our feet "roll" on the ground.

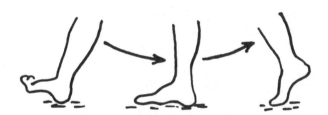

We come down to the ground with our heels and we lift the foot from our toes. When we run the heels come down only very lightly and swiftly — and we are far more on our toes than when we walk. It is as if when we run we want to give wings to our feet, like the Roman god Mercury had. In running we try to forsake the heaviness of the earth, and we are more on our toes than on our heels. But when soldiers march on a parade they want to give an impression of power and strength and their heels come down with force.

So there are two opposites in our foot: the heel which takes us down to earth, and the toes which we use to push away from the earth. You can see these two opposites in the bones of the foot. The bones of the heel are comparatively large and heavy, the bones of the toes are light. Like the vertebrae of the spine, the top ones are light and small and the lower ones are heavy and large — what goes up is light, what goes down is heavy.

When we run, we are more on our toes than on our heels, and for animals, running is far more important than it is for us. Whether an animal is a hunter or whether it is being hunted, running is of vital importance for it, and so most animals are on their toes all the time. For instance, a dog stands, walks and runs on four toes in each foot. The big toe is only a little stump

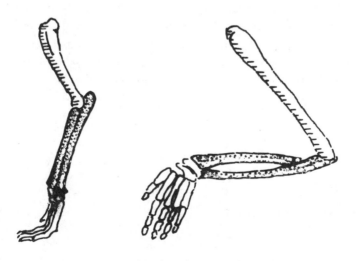

Forelimb of dog and man

and the heel never touches the ground. The horse's foot is even more interesting: the horse stands only on one toe, the middle toe which has grown very big, and its nail has grown over and around it becoming a hoof. The animals are all on their toes, their heels are high up and never come down except when they sit. This gives them greater speed than we have, but none of them can balance on two legs as we can.

Let us look at the muscles working in leg and foot. We first raise the heel at the beginning of a step using muscles in the calf at the back of the leg. Then the foot is lifted, it swings and comes down, and for this, the muscles in the front of the leg act mainly as a kind of brake so that the foot does not just drop down, but settles down gently.

There is a great contrast between these two sets of muscles. The muscles at the back must be stronger, for they have to lift the heel and with it the whole weight of the body resting on the heel. But the muscles in front only have to act as brakes to bring the movement gently to a stop. At the back of the leg are the muscles which start the movement and in front are those which end it.

There is one occasion when these front muscles of the leg are especially strained, and that is when you walk downhill. Then it takes much longer until the whole foot is down and so the muscles in front have a lot of extra work to do. That is why after a long downhill walk we can feel pain in the front of our leg. (The same thing happens in high-heeled shoes, when the high heel comes down too soon and the front muscles have extra work to do to let the toes come gently down.)

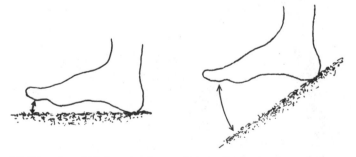

Walking downhill there is a longer distance for the toes to reach the ground, straining the muscles at the front of the leg

At the beginning of the step, when the heel is raised by the calf muscles, the weight of the body is shifted forwards. It no longer presses down on the strong bones of the heel, but on the smaller and looser bones forming the arch. The pressure is so great that the arch would collapse, dislocating the bones, if there were not muscles in the sole of the foot which keep the arch bent and the bones in place.

So we see now how many muscles are concerned with the rolling motion of the foot, the motion from heel to toe.

The Hand

Of all living creatures only we human beings have this complicated and different upright walk, and only the human body, the human skeleton is built in the right way for it. If you compare a human skeleton to the skeleton of a horse, you can see how heavy the horse looks beside the lightness of the human skeleton. Our very bones are built and arranged for upright walking.

But this upright walk also gives us something which no animal has. The animals stand on the ground in stable equilibrium with four legs. We human beings have to balance ourselves, keeping an unstable equilibrium but through this, our arms and hands are free. That is the great and wonderful advantage of the upright walk. It is a difficult, complicated motion — running on four legs would be much easier — but we gain from this difficult upright walk the free use of our hands.

At first sight the human hand is a very imperfect instrument. As a weapon the lion's paw is far more powerful than our hands. For digging, the mole's paws are far better than our hands. To hold fast, the bird's claw has a much stronger grip, or for flying a bird's wing is far superior to our hands. All these various animal limbs are specialized: for killing prey, for climbing, for scratching the earth, for fast running, while our hand is not specialized at all. It is the most primitive limb without any special skill built into it, but is so versatile it can create and make things. It can make weapons far more terrible than lions' claws, make tools which are better than any animal's paws, build machines, ships, aeroplanes, roads and cities. Above all else with our hands we can help each other. While each animal needs its forelegs for itself to carry the weight of the body, our hands can help and support others.

Scientists studying the embryos of animals have discovered a remarkable fact. At an early stage these tiny animal embryos in their mother's womb or in the egg have hands with five fingers like human hands, but as they grow, their hands change, becoming more specialized, changing into paws, claws, wings or whatever. So our hand is not the most advanced kind of limb, but the most primitive and the earliest forelimb. While we keep the original primitive hand, we can make tools which no animal can. Yet without the upright walk, this gift would be no use to us. The upright walk and the hand go together.

44

The Muscles

We looked at the work of some of the three hundred muscles, from the soles of our feet right up to our neck, which come into play with every single step we take. But there is also much other work done by the muscles. Just take a very simple occurrence. You see a friend, you smile, you say "hello," and you shake hands. You see your friend because muscles have turned your eyeball in their direction, other muscles have adjusted the lens in your eye so that you can see them clearly. You smile which involved muscles pulling up the corners of your lips. You speak and muscles stretch the vocal cords in the larynx in different ways. And when you shake hands muscles work in your arms and fingers. And while all this is going on, the muscles of your heart keep you alive by beating, the sheet of muscles of the diaphragm makes the lungs breathe, a very large group of muscles in the stomach digests your food and passes it on to other muscles of the intestines. There is no movement by the body and no movement inside the body which is not caused by the muscles.

It is of course the muscles which move the bones of your body, for the bones, stiff and hard as they are, cannot move themselves. To move anything in the world you need heat. We burn petrol to move cars or aeroplanes, we burn coal to move steam engines, and even when we use hydroelectric power, it was the heat of the sun which first raised the water. The muscles too can only work by producing heat. By far the largest part of the food we eat is carried by the blood from the intestines to the muscles, and is "burnt" in the muscles to provide the energy for all the countless movements. If people are starving the muscles become weaker and weaker and finally stop working altogether.

When we looked at bones, we saw the flow of water in the forming of bones. The story of the bones begins with water. But the story of the muscles begins with heat. Muscles have their strength from burning food. This burning or combustion is far superior to what happens in any kind of combustion engine we can build. There is some fantastically complicated chemistry which produces the "invisible flame" that gives our muscles strength. This invisible flame, this chemical combustion, wastes little material (while all our engines waste a very large part of the fuel) and produces energy while keeping a low temperature. No part of the body itself ever gets burned by this invisible flame while our petrol and diesel engines work at a very high temperature.

But this wonderful chemical burning has one thing in common with an ordinary flame, it leaves a kind of ash, lactic acid. The blood collects this ash and takes it away from the muscles. But if some muscles are made to work very hard without enough rest, then the blood can't clear the ash away fast enough. It accumulates, the muscles get stiff and sore, and we are forced to give them a long rest.

What do the muscles look like? The easy answer is, like raw meat. What people eat as meat is nothing but muscles. When we buy a cut of meat, the muscles are cut right through so we only see a cross-section of several muscles, not one muscle by itself. We can get some idea of what a single muscle looks like from the word "muscle." It comes from the Latin *musculus* meaning little mouse, for some muscles are shaped like this and it reminded the Romans of a mouse.

Each of these mouse-like shapes, each muscle, is composed of thousands and thousands of very fine but strong and elastic fibres. In one muscle, the biceps in our upper arm, there are about 600 000 of these strong elastic fibres.

How does a muscle actually work? As an example we will take the biceps in our upper arm. It is the muscle which draws or pulls the lower arm towards the upper arm every time we bend the arm. The biceps does this by contracting.

The muscle fibres are elastic and can contract or expand, but their real work is only done when they contract. The strength

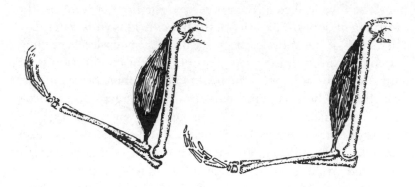

Action of the biceps muscle

of the biceps — and of every muscle in the body — lies only in the one direction of contracting. When the muscle relaxes and expands it has no strength whatsoever. So the biceps has no power to straighten the arm again. This has to be done by another muscle, the triceps at the back of the arm. As the triceps muscle contracts, it pulls the lower arm away from the upper, thus straightening the arm.

Of course, the biceps must relax and expand when its opposite, the triceps, pulls, and the triceps must relax when the biceps is pulling. But the two muscles are so marvellously coordinated that there is no conflict between them, and they do not hinder each other. So to move your lower arm you need two muscles working together in opposite directions. It is the same with any

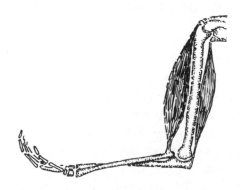

Relationship of biceps and triceps

muscle which moves any part of your body, there are always two of them working in pairs, one doing the opposite of the other. All the hundreds of muscles all over your body work in pairs.

A muscle works only when it contracts, but this contraction is by no means simple. Imagine you are carrying a bucket of water. The muscles of your arm have to contract as long as you carry the pail, but it is not all one long contraction. These muscles continually contract and expand a little about ten times in one second. Sometimes it can happen that through over-straining, a muscle cannot relax, cannot expand, and this is called muscle cramp.

There is one more mystery to which so far no answer has been found. Through practise and training in sport we can strengthen our muscles. What exactly happens in such specially trained muscles is still unknown: we know that it works, but we don't know how it works. It is one of the many still unsolved riddles of the human body.

Voluntary and Involuntary Muscles

We saw that muscles can only work by contracting, in other words, muscles can only pull the bones but cannot push them. We can move our limbs and fingers this way or that because there are always two muscles working in opposite directions, one pulling, its opposite relaxing. But whatever movement we make, it is never only one pair of muscles that comes into play. For a single step about three hundred muscles are called into action. When we lift something with our hand, dozens of muscles in our arms, shoulders, hips, even on our belly, are put to work. The muscles of the body are really like one great orchestra. In an orchestra sometimes one instrument takes the leading part and sometimes another, and so one muscle or another has the main task — like lifting something — and many other muscles accompany this. And all these different sets of muscles work in coordination or in harmony.

Some muscles bend and straighten, like the arm muscles; some muscles twist one way and back, like the muscles which turn the radius around the ulna; some muscles lift or lower, like the muscles in the leg, and some muscles slide in opposite directions. Yet all these different activities are in harmony so that the body moves and acts in unison.

The muscles of the body are divided into two large groups (with one important exception). The biceps, for instance, and its counterpart the triceps works only when you want them to work: they don't start bending your arm without your will to do so. They belong to the group of muscles which are called

   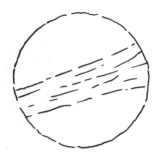

Voluntary muscle fibre (striped)   *Involuntary muscle fibre (smooth)*

voluntary muscles (*voluntas* is Latin for will). The intestines are also muscles, contracting and expanding and so pushing the food along. But you know nothing of this movement, you cannot command it or stop it. The intestines belong to the second group of muscles, the involuntary muscles.

Under a microscope these two groups of muscles look quite different. All muscles are composed of fibres, and a single fibre of a voluntary muscle is striped under a microscope. But the fibre of an involuntary muscle like the intestine is smooth. The smooth muscles do their job while we are quite unconscious of it and we cannot control them. We can only control the striped ones. The intestines and the stomach are smooth muscle fibres, while the muscle fibres in our arms and legs are striped.

The one exception is the muscles of the heart. We cannot control our heartbeat, we cannot command it to contract faster or slower, yet the heart muscles are striped. So in fact there are three types of muscle in the body: striped, smooth and cardiac (or heart) muscle. The heart muscle is in a category of its own, and there are other reasons for this. Although heart muscle is striped, the different fibres are joined together by branching which means the heart muscle beats really as one single entity. When it beats it is as if a rhythmic wave passes right through it. Also, of the three different kinds, heart muscle is easily the most active: our whole life long it can never take a moment's rest. And the heart muscle is more sensitive than the other two. For instance some people cannot drink coffee because it affects their

Cardiac muscle fibre (striped and branching)

heart too strongly. The heart is in fact sensitive to many things in and around us all the time whether we notice it or not.

Let us take a closer look at how a voluntary muscle like the biceps works. As we know, the biceps pull the lower arm towards the upper arm by contracting. To be able to pull, the muscle too must be firmly anchored on bone. This end of the muscle is called the origin of the muscle. The biceps has its origin in the shoulder blade. The other end of the muscle is attached to the bone which is pulled, in this case the radius, the lower arm. This other end is called insertion. Each voluntary muscle has an origin from where it pulls, and an insertion which is being pulled. Every muscle pulls towards the origin.

The insertion looks quite different from the origin. The muscle is soft, elastic, fleshy and full of blood vessels, but towards the insertion every voluntary muscle becomes a hard, tough cord which is almost as hard as bone. This tough cord is called tendon. Just above the heel you can feel the hard tendon of the calf muscle that has the task of lifting the heel and the weight of the body when you walk.

A muscle showing insertion, bulging flesh and tendon

The threefold human being consists of the head which is at rest, the limbs that move, and in between the rhythmical movement of heart and lungs. One part of us is at rest, another part moves, and a third is intermediary, moving rhythmically. It is the same with the muscle. At the origin the muscle is firmly anchored and at rest. At the insertion, the tough tendon pulls and moves the arm. But in the middle part, the soft, bulging flesh of the muscle contracts rhythmically. A muscle is not a little mouse as the Romans called it, but a little human being!

The task of any voluntary muscle is to pull a bone. And all this pulling is done by the principle of the lever. As far as the muscles are concerned, every bone in your body is a lever. In a lever we distinguish three parts: load, fulcrum and force. When you lift, let us say a ball, the hand with the ball is the load, the elbow joint is the fulcrum — and where the tendon of the biceps pulls is the force. Or take the moment when the heel of your foot is lifted: the muscle pulling up the heel is the force, the fulcrum is at the toes and the load is the weight pressing down through the shin bone and ankles upon the arch.

Even your ribs are used as levers. When the spine gets too far out of balance, certain muscles pull the ribs down in the opposite direction, using the ribs as levers to restore balance. In this case the fulcrum is in the hip, the load is the spine and the force pulls at the ribs. This only happens when you are far out of balance, normally the side processes of the vertebrae are used as levers by muscles in the back to keep the spine balanced.

So muscles and bones work together as a great system of levers and every time a muscle contracts, it turns a bone around a fulcrum to lift a load. It needs mechanics to understand how the muscles of the body work.

Human and Animal Skulls

The muscles are elastic, contracting and expanding, and they have this in common with the ligaments. But their task is quite different from that of the muscles. The muscles move the bones round the joints, using each joint as a fulcrum. But the ligaments are elastic bands, which hold the bones in their joints. When you lift your arm, a muscle pulls the humerus up and at the same time the ligament which keeps the head of the humerus in the joint is stretched. If there were no ligaments the bone would be dislocated, it would be wrenched from its joint with every pull of the muscles.

And when you lower your arm again, the ligament relaxes and contracts. The elasticity of the ligaments is the opposite of the elasticity of the muscles. The muscles work like a spring, you need force to make them contract and when the steel spring relaxes it expands. But the ligaments are like a rubber band. It needs force to stretch a rubber band and when you let go, the rubber band contracts. It is because the muscles and ligaments are elastic in opposite ways, that they work so wonderfully together enabling us to move our bones.

If a ligament is stretched too much, it gets inflamed and swollen. This is a sprain. A sprained ankle happens when the ligaments around the ankle bones have been stretched too much. They then have to rest before they can be used again. Even worse is a torn ligament which can take two or three months to heal.

Now let us look at the muscles in our neck. To understand what these muscles do, we must compare the skeleton of an animal, for instance a horse, with the human skeleton. The skull of

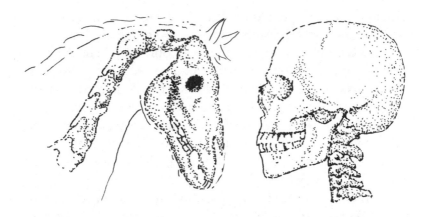

Animal skull and human skull

a horse is suspended from the neck, the head hangs on the spine just as you might hang a basket on a stick. It would be much harder to balance the basket on the stick, but that is exactly the position of the human head, the human skull. It is balanced on top of the spine; it is kept in unstable equilibrium on top of the spine and this balancing act is performed by the neck muscles.

When somebody falls asleep while sitting in a chair, their head drops to one side, because there is no longer the will in the neck muscles to balance the head on the top vertebrae. We not only balance the body on two slender supports, our legs, but we also balance the head on the spine. And this balance is not automatic. It is our own deed, our own will, and if we fall asleep we cannot perform this balancing. Think of a juggler balancing a ball on a finger: our neck muscles are such a juggler and they juggle the globe of the skull on the little vertebra, the atlas, on the top of the spine.

This is one difference between the human skull and an animal's skull. The animal's skull is suspended from the spine and the human skull is balanced on the spine. There is also another, more important difference between human and animal skulls. The shape of the human skull is more rounded, more like a sphere, while in comparison the animal's skull is long, like a cylinder. We found the cylinder shape in the shape of the limb bones, and the animal head is more like a limb. It is not only a matter of shape. As the animals use their forelegs to carry the body and are not free to use them for anything else, they use their heads for things we would do with our hands. The animals use their head as an extra limb. Think of a dog carrying a ball or a newspaper. Birds building their nests use their beaks, a part of their head, as we would use our hands. A lioness carries her cubs from one place to another in her mouth using her head as a limb. But the lion's mouth is also a weapon, its terrible fangs are as much a weapon as the claws of its paw. Bulls or stags use their heads when they fight.

So the animal skull resembles a limb not only in shape, but in use. By contrast the human head is set free from all these tasks; it has no physical work to perform, it is neither a tool, nor a weapon, and so it is free to think and to understand the world in a way which is not possible for the animals. The human head could develop a larger brain than an animal's just because the head is set free from physical work.

In a city there are shops, offices and factories — buildings where all the practical things and necessities are done. But there are also other buildings: those treated with reverence; buildings which you enter with a mood of respect: the churches, temples or chapels. In a church or temple no business is done, no things are produced, but if people enter these sacred buildings in the right spirit, they find something there that gives meaning to life. In the body our limbs do all the necessary and useful things and our heart, lungs and blood do all that is necessary to keep the body alive; they are like the shops and offices and factories. Our skull is the temple where no physical work is done, but where we give meaning and purpose to all we do and see.

The animal skull cannot be a temple as it must also work as a tool and as a weapon. But the human skull is the temple of the human spirit. Of all the bones, it is the most perfectly shaped. When you look at the skull, you should feel neither horror (as some people do) nor any kind of disrespect; you should feel that what you look at is a church or a temple of the human spirit that once lived in it.

The Shape of the Skull

Each bone of the human skeleton — the spine, the arms, the legs, the skull — can be described by itself, individually. But this would not be a true description because all these bones work together; they are like the individual letters forming a word. The letters H-O-U-S-E form the word house, and if you changed one single letter, the word would be different. Take the legs, for instance. We have looked in detail at the complicated work that goes on in the muscles and bones of the leg and foot when we take a single step. But if the legs were not built as they are, we would not be able to walk upright and our hands would not be free. If our hands were not free to do the things they do, we would not have our head free from physical work and it would not become the rounded dome it is.

So if each bone of the feet and the legs were not built as they are, we could not have the rounded dome which houses the brain. But the legs could not do their work if the spine were not built strong and flexible at the same time. And not only the spine, but also the pelvis of the human body is built to carry the centre of gravity of the body and to balance the weight.

The spine of a gorilla for instance is not as finely developed as ours: the vertebrae are not very different in size to one another and the spine cannot curve like ours. A gorilla's pelvis is larger and heavier and does not lie at the same angle to the spine. That's why apes do not have a true upright walk but lean forwards. Their skeleton is specialized for living in trees and swinging from branches. Their arms are longer and stronger

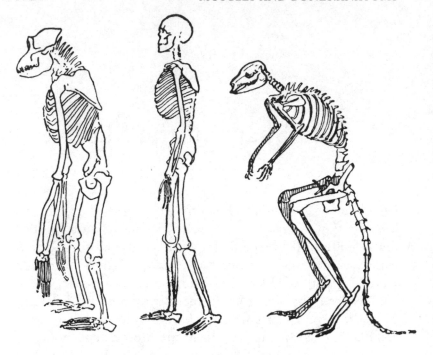

Skeleton of a gorilla, man and kangaroo

than their short legs. Their feet are used for grasping branches and are not made to carry the whole weight of the body. Only the human skeleton is built in every detail from head to foot for the upright walk.

As we saw, many animals use the head to perform tasks which we do with our hands, especially the task of grasping things. We grasp something by closing our fingers around it, but a dog or a cat has to grasp things with its mouth. Birds use their beaks, elephants, whose tusks have become pointed weapons, use their elongated nose, their trunk, to grasp things. Animals must grasp things with their heads. But in a certain way we human beings also grasp things with our heads, with our minds. We also use our head for grasping on a higher level — mentally not physically.

The animal has to grasp things physically with its jaws. It uses the jaws as we use the hands. In an animal's head the jaws come forward forming a snout while the forehead retreats. But in the human head it is the forehead which bulges forwards and the jaws which retreat. The animal has bulging jaws. In the apes, too, the jaws are far more prominent than the forehead. The skull of the ape is really a specialized tool — a great nut-cracker.

But what I told you about animals' heads is only true for the fully-grown animal. When we go back to the embryo stage, when the animal is still in its mother's womb, it is different. I told you already that at an early stage the animal embryo has tiny hands, like the human embryo — but that they change and become specialized tools. And something very similar happens with their head. At a certain early stage the animal embryo has a bulging forehead and small, retreating jaws — just like the human embryo — but as the embryo grows, the skull changes — the snout coming forward and the forehead receding.

This shows that the human being keeps the earlier shape of head while the animals develop to a later stage and lose the bulging forehead. Chimpanzees and apes keep the bulging forehead longer than other animals: a baby ape still has a bulging forehead when it is born, but loses it and develops the prominent jaw as it grows up. So the human skull with its

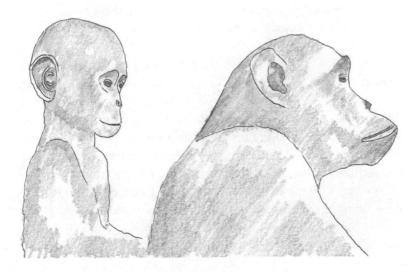

Young chimpanzee and adult chimpanzee

prominent forehead keeps the earliest shape while the animals change this shape.

Now think of the theory that the embryo in its development presents a picture of the ancestors of the growing child. This would mean that there was once a human-like creature from which not only *we* descend but also the animals. The animals changed while we present-day human beings kept more to the likeness of this human-like being. It would mean that the animals descend from man, not the other way around as the popular books put it. It is a possibility, and one that agrees with the development of animal embryos.

Looking at the skull itself — the temple of the human spirit — we see it consists of three parts. First there is the domelike rounded top part called the cranium. Secondly there are the bones forming the front of the skull, the face; these face bones have grown together with the cranium. And the third part is the jawbone which is quite separate and only jointed to the top.

The top part, the cranium, is the head in the real sense of the word; it is the shell which houses the brain. The bones in the middle part, which form the face, are much finer and lighter. The cheekbones curving forward remind you of the curve of the ribs. The third part, the jawbone, is the only bone of the head that can be moved like a limb. It is the only skull bone that is used for physical work when we chew our food and also when we speak.

This threefoldness of the skull becomes even clearer if we look at the functions. In physiology we had the nervous system, the rhythmical system and the digestive system. The nervous system has its centre in the brain, the rhythmical system has its centre in the ribcage, the heart and lungs, and the digestive system has its centre in the stomach and intestines. The top of the skull, the cranium, houses the brain, the centre of the nervous system. The middle part of the face has the opening of the nose taking in air for the lungs of the rhythmic system. In the lowest part of the skull, the upper and lower jaw together form the mouth, through which we take in food for the stomach and intestines.

The skull itself is again threefold. The cranium is the real skull, the face bones are similar to the ribs, and the jawbone that moves is a kind of limb bone. The whole skull is again an image of the whole human being.

Human Beings and Animals

The spine of the animal is parallel to the ground, lying horizontally. The human spine is at right angles to the earth and is vertical. Human spine and animal spine are at right angles to each other. Apes and kangaroos are animals which attempt a more upright position, but it is not a true right angle, for they lean forwards. It has to be so because neither the apes' nor the kangaroos' skeleton is really made for the difficult balancing act we perform when we stand or walk.

By far the greater number of animal families stand and walk with their spine in a horizontal position. And not only the spine but all the bones of the animal skeleton are built and arranged for a horizontal existence. Even the head of the animal is more horizontal. The projecting snout makes the animal skull more horizontal. The human skull on top of the vertical spine is rounded, but the arrangement of the three parts is vertical: the cranium rises above the middle part, the face bones, below which is the jawbone, the only jointed and movable bone of the head. In the horizontal animal head the jaw comes forward and the cranium recedes, so there is a horizontal rather than a vertical gesture.

The top part of the human skull forms a great cave containing the brain, surrounded and protected on all sides by strong bones. Like a castle built to keep out all enemies, the strong rounded bones let nothing of the world outside come near the brain. Farther down, where the face bones begin, there are two cavities which are open to the world, the eye sockets. Through the eyes all the light and colour, all the shapes of the world flow and stream in. At the back of each eye socket there is a small hole through which the nerves lead to the great inner cave of the brain.

All the light and colour of the world flows and streams into the eye, but only a fine transformed extract comes through this tiny hole to the inner cave of the brain. No light or colour comes to the brain directly, only the fine extract that passes along the optical nerves through the small hole at the back of the eye socket.

It is the same with the other cavities of the skull which open to the world outside. The hole of the nose is also an open cave with small openings for the nerves leading to the great inner cave of the brain. The mouth and the ear holes are the same.

The caves which open to the world outside — eyes, nose, mouth, ears — receive the light, colours, smells, tastes and sounds, but only a fine extract of these is allowed to come to the brain. That is why the skull so carefully protects the wonderful organ we call the brain.

Just as the Temple in Jerusalem had its Holy of Holies so the human skull has its Holy of Holies, the cavity in which the brain is contained and protected. I called it a great inner cave, but it is not like a cave in a mountain. When people built domes, they first built the walls which form the real church or temple, and over the vertical walls they constructed the vault of the ceiling, the dome. It is of course, the walls which carry and support the dome and form the enclosure of the temple. There are such vertical side-walls in our skull on either side of the cranium, and they are called the temples. The cave is really a dome, a temple with a vaulted ceiling.

But the skull, taken as a whole — cranium, face bones, jawbone — is an image of threefold man. The dome of the cranium contains the brain, the centre of the nervous system, the nose leads to the lungs which belong to the rhythmic system, and the mouth leads to the stomach, which belongs to the digestive system.

So we end these lessons about the anatomy as we began, with the threefoldness which can be found in all parts of the human body.

Index

Books by Charles Kovacs

Teacher resources

Class 4 (age 9–10)
Norse Mythology

Classes 4 and 5 (age 9–11)
The Human Being and the Animal World

Classes 5 and 6 (age 10–12)
Ancient Greece
Botany

Class 6 (age 11–12)
Ancient Rome

Classes 6 and 7 (age 11–13)
Geology and Astronomy

Class 7 (age 12–13)
The Age of Discovery

Classes 7 and 8 (age 12–14)
Muscles and Bones

Class 8 (age 13–14)
The Age of Revolution

Class 11 (age 16–17)
Parsifal and the Search for the Grail

Religion and spirituality

The Apocalypse in Rudolf Steiner's Lecture Series
Christianity and the Ancient Mysteries
The Michael Letters of Rudolf Steiner
The Spiritual Background to Christian Festivals

The Human Being and the Animal World

Charles Kovacs

An excellent comparative introduction to animals and humans for children, written in Charles Kovacs's warm and expressive style.

Charles Kovacs taught in Edinburgh so there is a Scottish flavour to the animals discussed in the first half of the book, including seals, red deer and eagles. In the later chapters, he covers elephants, horses and bears.

This is a fascinating and engaging resource for teachers and home schoolers which is used in the Steiner-Waldorf curriculum for 9–11 year olds.

florisbooks.co.uk

The Age of Discovery

Charles Kovacs

A wonderful overview of world history from the time of the Crusades to the Renaissance for children.

Including Saladin, Joan of Arc, Columbus, Magellan, Queen Elizabeth and Francis Drake, Kovacs explores the notion of cause and effect and the inter-connectedness of world events.

This is a fascinating and engaging resource for teachers and home schoolers which is used in the Steiner-Waldorf curriculum for 12–13 year olds.

florisbooks.co.uk

Botany

Charles Kovacs

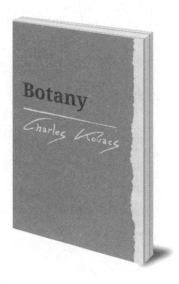

A wonderful introduction to botany for children written in Charles Kovacs's warm and expressive style.

Readers will explore what characterises different plants – from fungi, algae and lichens, to the lily and rose families – and learn about the parts of each plant and their growth cycle.

This is a fascinating and engaging resource for teachers and home schoolers which is used in the Steiner-Waldorf curriculum for 10–12 year olds..

florisbooks.co.uk

Norse Mythology

Charles Kovacs

An excellent retelling of the stories of Norse mythology for children which is written in Charles Kovacs's warm and expressive style.

Includes myths on Creation, Odin and Mimir, Thor and Thialfi, Idun, Sif and Loki.

This is a fascinating and engaging resource for teachers and home schoolers which is used in the Steiner-Waldorf curriculum for 9–10 year olds.

florisbooks.co.uk

Floris
Books

For news on all our **latest books**,
and to receive **exclusive discounts**,
join our mailing list at:

florisbooks.co.uk/signup

Plus subscribers get a FREE book
with every online order!